Industrial Engineering Strategy for Constructive Technologies

There is an urgent need to develop robust strategies to respond to and leverage new and emerging technologies, particularly those based on artificial intelligence (AI). Industrial engineering's systems-focused approach offers the best mechanism to address this urgent global need.

Industrial Engineering Strategy for Constructive Technologies: A Systems-Based Approach for the Global Economy focuses on managing digital engineering using a systems methodology to ensure that all the parts and pieces fit together. It addresses the role of AI, is cognizant of social concerns about technological encroachment, and highlights the sustainability of operations. This book leverages resilience engineering in technology utilization and, at the same time, recognizes humans in the loop of technology. This book also discusses how to use a systems-based approach for accepting and integrating new technologies. The global market is yearning for new guidelines and strategies for coping with the ever-increasing and changing technological landscape.

This book is an essential read for university students and instructors and those in the fields of engineering as well as industry, business, government, and the military.

Systems Innovation Book Series

Series Editor: Adedeji Badiru

Systems Innovation refers to all aspects of developing and deploying new technology, methodology, techniques, and best practices in advancing industrial production and economic development. This entails such topics as product design and development, entrepreneurship, global trade, environmental consciousness, operations and logistics, introduction and management of technology, collaborative system design, and product commercialization. Industrial innovation suggests breaking away from the traditional approaches to industrial production. It encourages the marriage of systems science, management principles, and technology implementation. Particular focus will be the impact of modern technology on industrial development and industrialization approaches, particularly for developing economics. The series will also cover how emerging technologies and entrepreneurship are essential for economic development and society advancement.

Project Management for the Oil and Gas Industry: A Global System Approach, Second Edition, 2-Volume Set
Adedeji B. Badiru

Project Management for the Oil and Gas Industry: Systems Engineering and Technology, Volume 1
Adedeji B. Badiru

Project Management for the Oil and Gas Industry: Agile Systems and Environmental Concerns, Volume 2
Adedeji B. Badiru

Handbook of Construction Management: Scope, Schedule, and Cost Control, Second Edition, 2-Volume Set
Edited by Abdul Razzak Rumane

Construction Management: Quality Tools and Techniques, Second Edition, Volume 1
Edited by Abdul Razzak Rumane

Construction Management: Project Management, Process Principles, and Concepts, Second Edition, Volume 2
Edited by Abdul Razzak Rumane

Industrial Engineering Strategy for Constructive Technologies: A Systems-Based Approach for the Global Economy
Adedeji B. Badiru

Industrial Engineering Strategy for Constructive Technologies

A Systems-Based Approach for the Global Economy

Adedeji B. Badiru

CRC Press
Taylor & Francis Group
Boca Raton New York London

CRC Press is an imprint of the
Taylor & Francis Group, an **informa** business

Designed cover image: Shutterstock

First edition published 2025
by CRC Press
2385 NW Executive Center Drive, Suite 320, Boca Raton FL 33431

and by CRC Press
4 Park Square, Milton Park, Abingdon, Oxon, OX14 4RN

CRC Press is an imprint of Taylor & Francis Group, LLC

ISBN: 978-1-032-65808-7 (hbk)
ISBN: 978-1-032-65921-3 (pbk)
ISBN: 978-1-032-65922-0 (ebk)

DOI: 10.1201/9781032659220

Typeset in Times
by SPi Technologies India Pvt Ltd (Straive)

Dedication

Dedicated to the memory of Omolade Bisade Badiru, the bud, who did not get a chance to blossom and contribute to the present age of constructive technologies.

Contents

Preface

This comprehensive solution-based book takes a constructive view of using new and emerging technologies to solve problems of the world from a systems-thinking perspective. The prevailing and persistent problems of the world are systems-based and must be tackled with systems-based methodologies. In contemporary utterances, businesses and industries claim to rely on disruptive technologies. This is to emphasize employing solution strategies that go against the status quo, going against the grain of normal practice. While "disruptive technologies" are recognized as a path forward, we believe highlighting such technologies as "constructive" is the way to go. Hence, the choice of "constructive technologies" in the book's title.

Two chosen platforms that convey the challenges around the world are the United Nations Sustainable Development Goals (SDGs) and the National Academy of Engineering's (NAE) 14 Grand Challenges for Engineering. This book uses these two platforms as the backdrop for proposing industrial engineering's systems-based approaches as recommended solution strategies. The world has not been idle in combating the noted challenges. However, progress reports have not indicated that we are where we need to be on the solution curve. Hence, our approach of leveraging the power of systems thinking within the industrial engineering framework is one additional alternative to pursue. Leveraging technologies requires recognizing the keystone role of humans in technological implementations. Industrial engineering is known for its people-centered approaches to problem-solving. Hence, focusing this book on using industrial engineering strategies makes practical sense.

The rapid growth of new and emerging technologies has created an urgent need for new strategies for operational efficiency, effectiveness, and productivity improvement. Digital engineering has emerged with new technological tools to provide new opportunities for how we live and work. The versatile and adaptive tools and techniques of industrial engineering provide a fertile avenue for responding to new and emerging technologies. From an operational perspective, the digital era consists primarily of digital-based science, technology, engineering, and mathematics. Learning to use the new developments requires rigorous and adaptive management processes. This book on managing digital engineering uses a systems methodology to ensure that all the parts and pieces fit together. Contents of the book include a multitude of contemporary topics, such as foundations of engineering in the society; emergence of digital engineering; using the DEJI systems model in digital engineering; educational platforms in the digital age; interfaces of teaching and research using digital platforms; managing diversity, equity, and inclusion in digital engineering; project management tools and techniques for digital engineering; achieving operational communication, cooperation, and coordination in digital engineering; developments in artificial intelligence, virtual engineering approaches, consistency, resilience, and responsiveness in the digital era; digital security in enterprises: business, industry, government, and the nation; people-in-the-loop safeguards in digital engineering; new development of COVID-response engineering; and project future developments in digital engineering. Of particular highlight in the book is the nexus of education, innovation, and invention. Education is essential for carrying everyone along on the digital journey. Without educational awareness, an undesirable room will be created for adverse conspiracy theories, similar to what was experienced in the early days of mobile telephones and what was pushed in some quarters about 5G technology. The management viewpoints advocated by this book will help pre-empt digital distractions so that we can maximize the output potentials of digital engineering and artificial intelligence.

The concluding chapter of this book presents the DEJI systems model. The model entails systems design, evaluation, justification, and integration. Systems quality is used as the basis for illustrating how the model is applied across its four stages. Specific emphasis is placed on applying the model to technology management in the context of pursuing and tackling the UN's SDGs and NAE's grand challenges.

Acknowledgments

We acknowledge and appreciate the support of our respective spouses and significant others throughout the arduous journey of writing this book and other related professional pursuits. As usual, the immensurable support, dedication, and expertise of the editorial team of Taylor & Francis Group are greatly appreciated. We thank Ms. Cindy Carelli, Ms. Kaitlyn Fisher, and their T&F colleagues for their bold and firm coordination of this book project from start to finish.

A coalition of engineering thoughts contributed and inspired me throughout the process of writing this book. These are individuals who use industrial engineering tools and techniques in their various respective organizations. I, hereby, acknowledge the motivational ideas of my daughter, Abi, a chemical engineer, who works as a process improvement engineer; my older son, Ade, a mechanical engineer, who works as a quality and performance enhancement engineer; my younger son, Tunji, a mechanical engineer, who works as an automation engineer; and my nephew, Femi, a mechanical engineer, who works as an energy research scientist. Their little doses of discussions led to a large collection of writing topics for this book. All their great ideas coalesced back to industrial engineering. All are recognized and appreciated.

About the Author

Adedeji B. Badiru is Dean Emeritus, Graduate School of Engineering and Management, USA Air Force Institute of Technology. While serving as dean, he had oversight for planning, directing, and controlling operations related to granting doctoral and master's degrees for the US Air Force. He was previously head of Systems Engineering with AFIT, head of Industrial Engineering with the University of Tennessee, and professor of Industrial Engineering and dean of the University College with the University of Oklahoma. He is currently a registered professional engineer, a certified project management professional, a fellow of the Institute of Industrial and Systems Engineers, a fellow of the Industrial Engineering and Operations Management Society, a fellow of the Institute for Operations Research and Management Science, and a fellow of the Nigerian Academy of Engineering. He is a prolific author and mentor.

Professor Badiru, a world-renowned educator, has been the recipient of several awards for teaching, research, and administration. He was a member of the AFIT team that won the US Air Force Organizational Excellence Award in 2019. He is the recipient of the Lifetime Achievement Award from Taylor & Francis Group in 2020, the Donald G. Newnan National Engineering Economy Teaching Excellence Award from the American Society for Engineering Education in 2024, and the Tau Beta Pi Engineering Honor Society Distinguished Alumnus award in 2024.

1 Technology and Systems in a Changing World

SYSTEMS BACKDROP

> While it is our duty to highlight what is wrong with the world, we also have an opportunity to point out what is going right, what is possible and how the United Nations is making a difference.
>
> – **Melissa Fleming**, Under-Secretary-General for Global Communications

The abovementioned quote is representative of the author's resolve not only to identify deficiencies but also to offer constructive systems-based solution strategies. This comprehensive solution-based book takes a constructive view of using new and emerging technologies to solve problems of the world from a systems-thinking perspective. The prevailing and persistent problems of the world are systems-based and must be tackled with systems-based methodologies. In contemporary utterances, businesses and industries claim to rely on disruptive technologies. This is to emphasize employing solution strategies that go against the status quo, going against the grain of normal practice. While "disruptive technologies" are recognized as a path forward, we believe highlighting such technologies as "constructive" is the way to go. Hence, the choice of "constructive technologies" in this book's title.

> It is not the strongest of the species that survives, nor the most intelligent that survives. It is the one that is most adaptable to change.
>
> – **Charles Darwin**

Technology brings about changes, and it facilitates pathways for adapting to the changes. For the world to survive and advance, technology must be managed to play its role. The sweeping changes now being driven by artificial intelligence (AI) systems is a good example of the world's adaptability to change. Technology has revolutionized the way we conduct business around the world. Integrating technology into our prevailing social and human infrastructure is the recommended path forward. The position of this book is that industrial engineering tools and techniques are critical for achieving the desired goal. The integration template provided by the Design, Evaluation, Justification, Integration (DEJI) systems model (Badiru, 2023c) could be an effective tool for this purpose. A good case example of a failure to adapt to a fast-changing technology platform is the case of the demise of Blackberry mobile phones. Blackberry fell precipitously out of the smartphone market because they failed to adapt and enmesh to the changed technology platform presented by Apple, which was a highly vaunted competitor. In 2016, Blackberry was still optimistic that

DOI: 10.1201/9781032659220-1

they could surmount the Apple challenge with their own BB10 operating system. Unfortunately, they missed the road completely. A movie was made to depict the rapid rise and fatal fall of the initially popular Blackberry smartphones. It was like the fall of the Roman empire, brought on by rapid changes around the empire.

Two chosen platforms that convey the contemporary challenges around our present world are the United Nations (UN) Sustainable Development Goals (SDGs) and the National Academy of Engineering's (NAE) 14 Grand Challenges for Engineering. This book uses these two platforms as the backdrop for proposing industrial engineering's systems-based approaches as recommended solution strategies. The world has not been idle in combating the noted challenges. However, progress reports have not indicated that we are where we need to be on the solution curve. Hence, our approach of leveraging the power of systems-thinking within the industrial engineering framework is one additional alternative to pursue. Leveraging technologies requires recognizing the keystone role of humans in technological implementations. Industrial engineering is known for its people-centered approaches to problem-solving. Hence, focusing this book on using industrial engineering strategies makes practical sense. This book is written by a team of engineers with diverse technical and managerial backgrounds. Thus, the flavoring of the book is comprehensive and inclusive. The quotes given below articulate the value of engineering in tackling the challenges of the world.

> Scientists study the world as it is, engineers create the world that never has been.
>
> – **Dr. Theodore Von Karman**

> If you list the top problems that America has to deal with – the environment, energy, national security, homeland security, the economy, water supplies, and health care, the answers are going to come from engineers, not lawyers, bankers, or accountants.
>
> **Norm Augustine** (former CEO of Lockheed Martin, in Aviation Week and Space Technology, May 11, 2009)

Very bold statements but true. Through their versatile tools and techniques, industrial engineers work on the products of science and engineering to make the world a better place for all.

Engineers are known for making stuff. As Elon Musk quoted:

> If you don't make stuff, there is no stuff.

So, here we have it. Engineers make stuff while industrial engineers make stuff better.

EMERGENCE OF DISRUPTIVE TECHNOLOGIES

Although the title of this book advocates "constructive technologies" rather than "disruptive technologies," the reality of today is that disruptive technologies are needed to help change our conventional practices of laissez-faire, acceptance, and business as usual. The view of this book is that the avowed disruptive technologies are actually constructive in terms of what we can accomplish positively with them. No new technology is expected to cause disruption of organizational missions. Rather, any new technology is expected to construct new paradigms that move organizations progressively toward their strategies, goals, and objectives. A ***disruptive technology*** is defined as:

> An innovation that significantly alters the way consumers, industries, and businesses operate.

Disruptive technologies introduce rapid changes that can render existing products and systems obsolete. Thus, forcing new ways of thinking and operating. Unlike sustaining technologies that offer continuous and incremental improvements, disruptive technologies bring about (or force) rapid changes and adaptations, possibly, in unexpected ways. In recent years, the most prevalent disruptive technologies include the following:

1. AI
2. Quantum science
3. Quantum computing
4. Virtual reality
5. Augmented reality
6. Data analytics and fusion
7. Cybersecurity models
8. Autonomous systems
9. Human-agent teaming
10. Smart materials
11. Blockchain technologies
12. Remote sensing
13. Digitization
14. Chat Bots and Chat-GPT
15. Internet of Things
16. 5G technology
17. Lithium-metal energy packs
18. Greening systems
19. Electric vehicles
20. Solar energy devices
21. Ultra smartphones
22. Video streaming
23. Audio streaming
24. Social media platforms
25. CRISPR medical technology (clustered regularly interspaced short palindromic repeats)

It is interesting to note that in its early days, television was seen (and ignored) as an emerging disruptive technology that would not never take hold enough to compete with radio.

Badiru, Ciarallo, and Mbonimpa (2023) present some landmark publications from the US Air Force Institute of Technology (AFIT) that address some critical implementations of technologies in our societal systems, both military and civilian. The table of contents include the following selections:

- Evaluating information assurance strategies
- Analysis of systems security engineering design principles for the development of secure and resilient systems
- Operator suspicion and human-machine team performance under mission scenarios of unmanned ground vehicle operation
- Performance evaluations of quantum key distribution system architectures
- Multimodal data fusion for systems improvement: A review
- Cost estimating using a new learning curve theory for non-constant production rates
- A learning curve model accounting for the flattening effect in production cycles
- Shifting satellite control paradigms: Operational cybersecurity in the age of megaconstellations
- Using a CubeSat reference architecture for accelerated model development and analysis
- Cislunar debris propagation following a catastrophic spacecraft mishap
- Elliptical orbit proximity operations differential games
- Systems modeling language extension to support modeling of human-agent teams
- Effects of agent timing on the human-agent team
- Spatialized audio improves call sign recognition during multi-aircraft control
- Method for generating a figure of merit for the selection of antennas under volumetric constraint
- First principle scintillation characterization of natural and artificial disturbances on V/W band signals in the ionosphere using the multiple phase screen technique
- First principle EMI model of wideband signal temporal delay induced by a High Altitude Nuclear Explosion (HANE) in the ionosphere
- Wideband SATCOM model: Evaluation of numerical accuracy and efficiency
- Understanding flow characteristics in metal additive manufacturing
- Fatigue life modeling and experimentation of additively manufactured components with respect to defect size and location
- Analysis and simulation of hypervelocity gouging impacts for a high-speed sled test
- Nonlinear dynamic analysis of an icosahedron frame which exhibits chaotic behavior
- Quantitative analysis of cerium-gallium alloys using a hand-held laser-induced breakdown spectroscopy device
- Measurement of electron density and temperature from laser-induced nitrogen plasma at elevated pressure (1–6 bar)

- On-chip silicon photonic controllable 2×2 four-mode waveguide switch
- Three-dimensional Fabry–Perot cavities sculpted on fiber tips using a multiphoton polymerization process

The premise of this book is that any of these disruptive (albeit constructive) technologies are best integrated into our existing operating scenarios, using the human-centered tools and techniques of industrial engineering, of which systems integration is both the cornerstone and keystone of operations. Another chapter of this book presents the DEJI systems model for technology and systems integration. The model presents a structural template for systems design, evaluation, justification, and integration. Systems quality is used as the basis for illustrating how the model is applied across its four stages. Specific emphasis is placed on applying the model to technology management in the context of pursuing and tackling the UN's SDGs and the NAE's grand challenges.

THE UNITED NATIONS SYSTEM

The UN, whose functions cut across geographical and functional boundaries across the world, is, indeed, a system. It is a behemoth organization but still a system with tentacles reaching far corners of the world. The complexity of world problems confirms that we live in a systems world. Solutions that are regimented and segmented on local platforms will not work in the long run. In the authors' extensive research, the solutions achieved don't fully represent the level of efforts, time, and resources committed to those solutions. This book aims to advance the solution process further by calling upon the systems-based strategies of industrial engineering. Many approaches have been tried, and many are being explored daily. The approach of this book adds to the body of knowledge available to the world in tackling world problems, each and every subsystem at a time, in coordination with other subsystems. Figure 1.1 presents the official chart representing the inner workings of the UN. As can be seen in the figure, every conceivable element of the human-based world is addressed somewhere in the chart, even if names and terminologies are different.

THE 17 SDGs OF THE UN

In 2015, seeking to partition world problems into manageable chunks, the UN created the 17 SDGs. At that time, world leaders came together and made a historic promise to secure the rights and well-being of everyone on a healthy and thriving planet. The result was the creation of the 2030 Agenda for sustainable development and the associated 17 SDGs. The authors here believe that adding industrial engineering tools and techniques to the mix of solution strategies will get the world further along toward the 2030 yardstick.

The 2030 agenda provides a shared blueprint for peace and prosperity for people and the planet, now and into the future. At its heart are the 17 SDGs, which are an urgent call for action by all countries – developed and developing – in a global partnership. They recognize that ending poverty and other deprivations must go hand-in-hand with strategies that improve health and education, reduce inequality, and spur

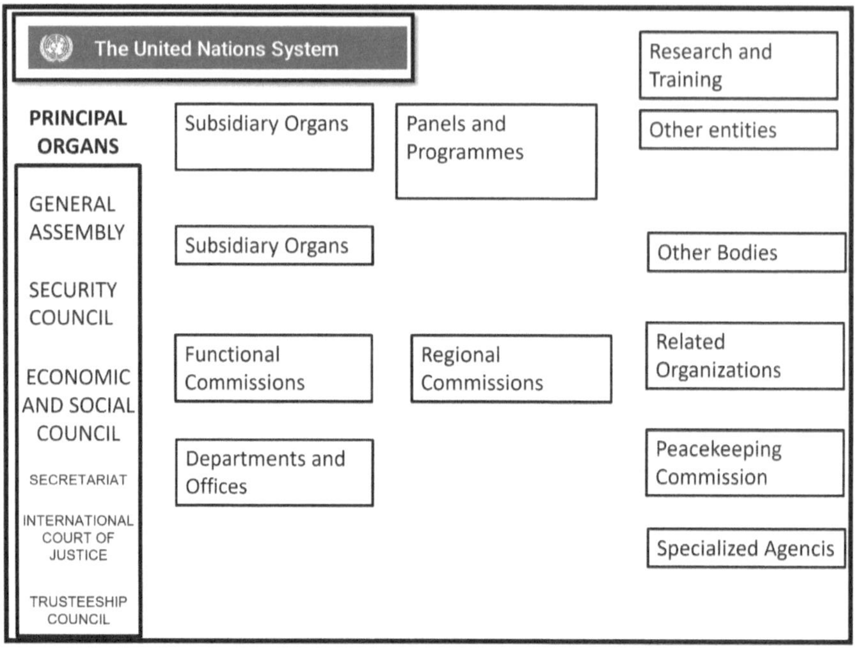

FIGURE 1.1 The United Nations as a System.

Source: https://www.un.org/sites/un2.un.org/files/un_system_chart.pdf. Reprinted with attribution and permission: https://www.un.org/en/department-global-communications

economic growth – all while tackling climate change and working to preserve oceans and forests. The SDGs build on decades of work by countries and the UN, including the UN Department of Economic and Social Affairs. Although the SDGs appear to focus more on human welfare, they are, indeed, pertinent and of interest to business, industry, academia, government, and the military. This is because whatever affects people, ultimately, will affect the interests of those national segments.

- In June 1992, at the Earth Summit in Rio de Janeiro, Brazil, more than 178 countries adopted Agenda 21, a comprehensive plan of action to build a global partnership for sustainable development to improve human lives and protect the environment.
- Member States unanimously adopted the Millennium Declaration at the Millennium Summit in September 2000 at the UN Headquarters in New York. The Summit led to the elaboration of eight Millennium Development Goals (MDGs) to reduce extreme poverty by 2015.
- The Johannesburg Declaration on Sustainable Development and the Plan of Implementation, adopted at the World Summit on Sustainable Development in South Africa in 2002, reaffirmed the global community's commitments to poverty eradication and the environment and built on Agenda 21 and the Millennium Declaration by including more emphasis on multilateral partnerships.

- At the United Nations Conference on Sustainable Development (Rio+20) in Rio de Janeiro, Brazil, in June 2012, Member States adopted the outcome document "The Future We Want" in which they decided, inter alia, to launch a process to develop a set of SDGs to build upon the MDGs and to establish the UN High-level Political Forum on Sustainable Development. The Rio+-20 outcome also contained other measures for implementing sustainable development, including mandates for future programs of work in development financing, small island developing states, and more.
- In 2013, the General Assembly set up a 30-member Open Working Group to develop a proposal on the SDGs.
- In January 2015, the General Assembly began the negotiation process on the post-2015 development agenda. The process culminated in the subsequent adoption of the 2030 Agenda for Sustainable Development, with 17 SDGs at its core, at the UN Sustainable Development Summit in September 2015.
- 2015 was a landmark year for multilateralism and international policy shaping, with the adoption of several major agreements:
  - Sendai Framework for Disaster Risk Reduction (March 2015).
  - Addis Ababa Action Agenda on Financing for Development (July 2015).
  - Transforming our world: the 2030 Agenda for Sustainable Development with its 17 SDGs was adopted at the UN Sustainable Development Summit in New York in September 2015.
  - Paris Agreement on Climate Change (December 2015).
- Now, the annual High-level Political Forum on Sustainable Development serves as the central UN platform for the follow-up and review of the SDGs.

Today, the Division for Sustainable Development Goals (DSDG) in the United Nations Department of Economic and Social Affairs (UNDESA) provides substantive support and capacity-building for the SDGs and their related thematic issues, including water, energy, climate, oceans, urbanization, transport, science and technology, the Global Sustainable Development Report (GSDR), partnerships and small island developing states. DSDG plays a key role in the evaluation of the UN system-wide implementation of the 2030 Agenda and on advocacy and outreach activities relating to the SDGs. In order to make the 2030 Agenda a reality, broad ownership of the SDGs must translate into a strong commitment by all stakeholders to implement the global goals. DSDG aims to help facilitate this engagement.

The 17 SDGs are presented below. The 2024 progress report on the SDGs is presented in Appendix A.

Goal 1: End Poverty in All Its Forms Everywhere

Despite considerable progress in the fight against poverty since 1990, over 800 million people, 70% of whom are women, still live in extreme poverty. The new Sustainable Development Agenda will now aim to eradicate extreme poverty by 2030.

Besides aiming to eradicate extreme poverty, Goal 1 takes a comprehensive approach to poverty as a whole by including a target on relative poverty based on national definitions. The poor are particularly vulnerable to economic and political crisis, loss of biodiversity and ecosystem services, and natural disasters and violence. To ensure that people who have escaped poverty do not fall back into it, this goal also envisages measures to strengthen livelihood resilience, including the establishment of social security systems.

1.1: By 2030, eradicate extreme poverty for all people everywhere, currently measured as people living on less than $1.25 a day
1.2: By 2030, reduce at least by half the proportion of men, women, and children of all ages living in poverty in all its dimensions according to national definitions
1.3: Implement nationally appropriate social protection systems and measures for all, including floors, and by 2030, achieve substantial coverage of the poor and the vulnerable
1.4: By 2030, ensure that all men and women, in particular the poor and the vulnerable, have equal rights to economic resources, as well as access to basic services, ownership and control over land and other forms of property, inheritance, natural resources, appropriate new technology, and financial services, including microfinance
1.5: By 2030, build the resilience of the poor and those in vulnerable situations and reduce their exposure and vulnerability to climate-related extreme events and other economic, social, and environmental shocks and disasters
1.a: Ensure significant mobilization of resources from a variety of sources, including through enhanced development cooperation, in order to provide adequate and predictable means for developing countries, in particular least developed countries, to implement programs and policies to end poverty in all its dimensions
1.b: Create sound policy frameworks at the national, regional, and international levels, based on pro-poor and gender-sensitive development strategies, to support accelerated investment in poverty eradication actions

Goal 2: End Hunger, Achieve Food Security and Improved Nutrition, and Promote Sustainable Agriculture

Although the situation has improved in numerous countries, many people still suffer from hunger and malnutrition around the world. Undernourishment affects nearly 800 million people worldwide – most of them women and children. The aim of the 2030 Agenda is to end hunger and all forms of malnutrition around the world within the next 15 years. In view of the rapidly growing global demand for food, it is estimated that food production worldwide will need to have doubled by 2050. Some 70% of people who are undernourished rely directly or indirectly on agriculture for their livelihoods; hence, it is smallholder farmers who are at risk of undernourishment.

Besides aiming to eliminate hunger, Goal 2 includes a target on ending all forms of malnutrition. Food quality is just as important as food quantity. Goal 2 also incorporates economic aspects, such as doubling the agricultural productivity and income of small-scale farmers by 2030. Furthermore, it features provisions on sustainable agriculture to prevent increased food production from harming the environment.

2.1: By 2030, end hunger and ensure access by all people, in particular the poor and people in vulnerable situations, including infants, to safe, nutritious, and sufficient food all year round
2.2: By 2030, end all forms of malnutrition, achieving, by 2025, the internationally agreed targets on stunting and wasting in children under 5 years of age, and address the nutritional needs of adolescent girls, pregnant and lactating women, and older persons
2.3: By 2030, double the agricultural productivity and incomes of small-scale food producers, in particular women, indigenous peoples, family farmers, pastoralists, and fishers, including through secure and equal access to land, other productive resources and inputs, knowledge, financial services, markets and opportunities for value addition, and non-farm employment
2.4: By 2030, ensure sustainable food production systems and implement resilient agricultural practices that increase productivity and production; that help maintain ecosystems; that strengthen capacity for adaptation to climate change, extreme weather, drought, flooding, and other disasters; and that progressively improve land and soil quality
2.5: By 2020, maintain the genetic diversity of seeds, cultivated plants, and farmed and domesticated animals and their related wild species, including through soundly managed and diversified seed and plant banks at the national, regional, and international levels, and promote access to and fair and equitable sharing of benefits arising from the utilization of genetic resources and associated traditional knowledge, as internationally agreed
2.a: Increase investment, including through enhanced international cooperation, in rural infrastructure, agricultural research and extension services, technology development, and plant and livestock gene banks in order to enhance agricultural productive capacity in developing countries, in particular, least developed countries
2.b: Correct and prevent trade restrictions and distortions in world agricultural markets, including through the parallel elimination of all forms of agricultural export subsidies and all export measures with equivalent effect, in accordance with the mandate of the Doha Development Round
2.c: Adopt measures to ensure the proper functioning of food commodity markets and their derivatives and facilitate timely access to market information, including food reserves, in order to help limit extreme food price volatility

Goal 3: Ensure Healthy Lives and Promote Well-Being for All at All Ages

The MDGs have made a significant contribution to improving global health, for example, in the fight against diseases such as AIDS, tuberculosis, and malaria. For example, the number of malaria deaths has fallen by 60% since 2000. However, results have failed to meet expectations in many areas, such as reducing child and maternal mortality.

The experience of the MDGs has shown that health problems need to be viewed in context, not in isolation. Education and food security influence the effectiveness of healthcare programs. Goal 3 continues along the same lines as the MDGs, for example with regard to child and maternal mortality, as well as communicable diseases such as AIDS, malaria, and tuberculosis, while also including provisions on combating non-communicable diseases such as diabetes and preventing traffic accidents and narcotics abuse. All people should have access to good-quality healthcare and medicines, including financial risk protection. Another objective for 2030 is to ensure universal access to sexual and reproductive healthcare, including family planning, information, and education.

3.1: By 2030, reduce the global maternal mortality ratio to less than 70 per 100,000 live births

3.2: By 2030, end preventable deaths of newborns and children under 5 years of age, with all countries aiming to reduce neonatal mortality to at least as low as 12 per 1,000 live births and under-5 mortality to at least as low as 25 per 1,000 live births

3.3: By 2030, end the epidemics of AIDS, tuberculosis, malaria, and neglected tropical diseases and combat hepatitis, water-borne diseases, and other communicable diseases

3.4: By 2030, reduce by one-third premature mortality from non-communicable diseases through prevention and treatment and promote mental health and well-being

3.5: Strengthen the prevention and treatment of substance abuse, including narcotic drug abuse and harmful use of alcohol

3.6: By 2020, halve the number of global deaths and injuries from road traffic accidents

3.7: By 2030, ensure universal access to sexual and reproductive healthcare services, including for family planning, information, and education, and the integration of reproductive health into national strategies and programs

3.8: Achieve universal health coverage, including financial risk protection, access to quality essential healthcare services, and access to safe, effective, quality, and affordable essential medicines and vaccines for all

3.9: By 2030, substantially reduce the number of deaths and illnesses from hazardous chemicals and air, water, and soil pollution and contamination

3.a: Strengthen the implementation of the World Health Organization Framework Convention on Tobacco Control in all countries, as appropriate

3.b: Support the research and development of vaccines and medicines for the communicable and non-communicable diseases that primarily affect developing countries, provide access to affordable essential medicines and vaccines, in accordance with the Doha Declaration on the TRIPS Agreement and Public Health, which affirms the right of developing countries to use to the full the provisions in the Agreement on Trade-Related Aspects of Intellectual Property Rights regarding flexibilities to protect public health, and, in particular, provide access to medicines for all

3.c: Substantially increase health financing and the recruitment, development, training, and retention of the health workforce in developing countries, especially in least developed countries and small island developing states
3.d: Strengthen the capacity of all countries, in particular developing countries, for early warning, risk reduction, and management of national and global health risks

Goal 4: Ensure Inclusive and Equitable Quality Education and Promote Lifelong Learning Opportunities for All

The international community has reaffirmed the importance of education and good-quality training in improving the living conditions of individuals, communities, and entire societies. Drawing many useful lessons from the MDGs, the new sustainable development Goal 4 goes beyond children's primary education, highlighting in particular the link between basic education and vocational training. Furthermore, it emphasizes equity and quality of education in a lifelong learning approach, two aspects which were not addressed in the MDGs.

Goal 4 aims to ensure that all children, adolescents, and adults – especially those most marginalized and vulnerable – have access to education and training appropriate to their needs and the context in which they live. This makes education a contributing factor in making the world more secure, sustainable, and interdependent.

Switzerland has comparative advantages with its systems of vocational training (dual vocational education and training) and basic education (decentralized governance, bridges, bilingualism). Through the support it provides, Switzerland improves access to good-quality education for all and strengthens the links between basic education, vocational training, and socio-occupational integration in order to boost economic and social development.

4.1: By 2030, ensure that all girls and boys complete free, equitable, and quality primary and secondary education leading to relevant and effective learning outcomes
4.2: By 2030, ensure that all girls and boys have access to quality early childhood development, care, and pre-primary education so that they are ready for primary education
4.3: By 2030, ensure equal access for all women and men to affordable and quality technical, vocational, and tertiary education, including university
4.4: By 2030, substantially increase the number of youth and adults who have relevant skills, including technical and vocational skills, for employment, decent jobs, and entrepreneurship
4.5: By 2030, eliminate gender disparities in education and ensure equal access to all levels of education and vocational training for the vulnerable, including persons with disabilities, indigenous peoples, and children in vulnerable situations
4.6: By 2030, ensure that all youth and a substantial proportion of adults, both men and women, achieve literacy and numeracy

4.7: By 2030, ensure that all learners acquire the knowledge and skills needed to promote sustainable development, including, among others, through education for sustainable development and sustainable lifestyles, human rights, gender equality, promotion of a culture of peace and non-violence, global citizenship, and appreciation of cultural diversity and of culture's contribution to sustainable development

4.a: Build and upgrade education facilities that are child, disability, and gender sensitive and provide safe, non-violent, inclusive, and effective learning environments for all

4.b: By 2020, substantially expand globally the number of scholarships available to developing countries, in particular least developed countries, small island developing states and African countries, for enrollment in higher education, including vocational training and information and communications technology, technical, engineering, and scientific programs, in developed countries and other developing countries

4.c: By 2030, substantially increase the supply of qualified teachers, including through international cooperation for teacher training in developing countries, especially least developed countries and small island developing states

Goal 5: Achieve Gender Equality and Empower All Women and Girls

Gender inequality is one of the biggest obstacles to sustainable development, economic growth, and poverty reduction. Thanks to MDG 3 on gender equality and women's empowerment, progress in enrolling girls into school and integrating women into the job market has been considerable. MDG 3 gave the issue of gender parity a lot of visibility, but its narrow focus meant that it failed to address important issues such as violence against women, economic disparities, and the low participation of women in political decision-making.

Goal 5 advocates equal opportunities for men and women in economic life, the elimination of all forms of violence against women and girls, the elimination of early and forced marriage, and equal participation at all levels.

5.1: End all forms of discrimination against all women and girls everywhere

5.2: Eliminate all forms of violence against all women and girls in the public and private spheres, including trafficking and sexual and other types of exploitation

5.3: Eliminate all harmful practices, such as child, early, and forced marriage and female genital mutilation

5.4: Recognize and value unpaid care and domestic work through the provision of public services, infrastructure and social protection policies, and the promotion of shared responsibility within the household and the family as nationally appropriate

5.5: Ensure women's full and effective participation and equal opportunities for leadership at all levels of decision-making in political, economic, and public life

5.6: Ensure universal access to sexual and reproductive health and reproductive rights as agreed in accordance with the Program of Action of the International Conference on Population and Development and the Beijing Platform for Action and the outcome documents of their review conferences

5.a: Undertake reforms to give women equal rights to economic resources, as well as access to ownership and control over land and other forms of property, financial services, inheritance, and natural resources, in accordance with national laws

5.b: Enhance the use of enabling technology, in particular information and communications technology, to promote the empowerment of women

5.c: Adopt and strengthen sound policies and enforceable legislation for the promotion of gender equality and the empowerment of all women and girls at all levels

Goal 6: Ensure Availability and Sustainable Management of Water and Sanitation for All

Access to drinking water and sanitation is a human right, and together with water resources, a key determinant in all aspects of social, economic, and environmental development. The MDGs incorporated targets on drinking water and sanitation but did not address other aspects crucial to sustainable development, such as water resources management, wastewater management, water quality, and enhancing resilience to water-related disasters.

Besides access to drinking water, sanitation, and hygiene, Goal 6 thus comprises additional targets such as protecting and restoring water-related ecosystems (including mountains, forests, wetlands, rivers, and lakes). Goal 6 aims to improve water quality and reduce water pollution, especially the type caused by hazardous chemicals. It also advocates cross-border cooperation as the key to managing water resources in an integrated fashion at all levels.

6.1: By 2030, achieve universal and equitable access to safe and affordable drinking water for all

6.2: By 2030, achieve access to adequate and equitable sanitation and hygiene for all and end open defecation, paying special attention to the needs of women and girls and those in vulnerable situations

6.3: By 2030, improve water quality by reducing pollution, eliminating dumping, and minimizing release of hazardous chemicals and materials, halving the proportion of untreated wastewater and substantially increasing recycling and safe reuse globally

6.4: By 2030, substantially increase water-use efficiency across all sectors and ensure sustainable withdrawals and supply of freshwater to address water scarcity and substantially reduce the number of people suffering from water scarcity

6.5: By 2030, implement integrated water resources management at all levels, including through transboundary cooperation as appropriate

6.6: By 2020, protect and restore water-related ecosystems, including mountains, forests, wetlands, rivers, aquifers, and lakes
6.a: By 2030, expand international cooperation and capacity-building support to developing countries in water- and sanitation-related activities and programs, including water harvesting, desalination, water efficiency, wastewater treatment, recycling, and reuse technologies
6.b: Support and strengthen the participation of local communities in improving water and sanitation management

Goal 7: Ensure Access to Affordable, Reliable, Sustainable, and Modern Energy for All

Access to energy is an essential prerequisite to achieving many SDGs that extend far beyond the energy sector, such as eradicating poverty, increasing food production, providing clean water, improving public health, enhancing education, creating economic opportunity, and empowering women. At present, 1.6 billion people worldwide have no access to electricity, while 2.5 billion rely on traditional biomass as a source of energy.

Goal 7 thus advocates universal access to affordable, reliable, and modern energy services. Given that sustainable development hinges on climate-friendly economic development, Goal 7 aims for a substantial increase in the share of renewables in the global energy mix and a doubling in the global rate of improvement in energy efficiency. Another target is to promote research in renewable energy and energy efficiency, as well as investment in energy infrastructure and clean energy technologies.

Goal 7: Ensure access to affordable, reliable, sustainable, and modern energy for all

7.1: By 2030, ensure universal access to affordable, reliable, and modern energy services
7.2: By 2030, increase substantially the share of renewable energy in the global energy mix
7.3: By 2030, double the global rate of improvement in energy efficiency
7.a: By 2030, enhance international cooperation to facilitate access to clean energy research and technology, including renewable energy, energy efficiency, and advanced and cleaner fossil-fuel technology, and promote investment in energy infrastructure and clean energy technology
7.b: By 2030, expand infrastructure and upgrade technology for supplying modern and sustainable energy services for all in developing countries, in particular least developed countries, small island developing states and landlocked developing countries, in accordance with their respective programs of support

Goal 8: Promote Sustained, Inclusive, and Sustainable Economic Growth, Full and Productive Employment and Decent Work For All

According to current figures, over 200 million people around the world are unemployed, especially young people. Employment and economic growth play a significant part in combating poverty. The promotion of sustainable growth and green business practices and the creation of sufficient decent jobs, in a context of respect for human rights and planetary boundaries, are of key importance not only for developing countries but for emerging and industrialized economies too.

Goal 8 includes targets on sustaining economic growth, increasing economic productivity, and creating decent jobs. It envisages combating forced labor and ending modern slavery and human trafficking by 2030. Sustainable economic growth should not be detrimental to the environment, which is why Goal 8 also advocates improved global resource efficiency in consumption and production and the decoupling of economic growth from environmental degradation.

8.1: Sustain per capita economic growth in accordance with national circumstances and, in particular at least 7 percent gross domestic product growth per annum in the least developed countries

8.2: Achieve higher levels of economic productivity through diversification, technological upgrading, and innovation, including through a focus on high value-added and labor-intensive sectors

8.3: Promote development-oriented policies that support productive activities, decent job creation, entrepreneurship, creativity, and innovation and encourage the formalization and growth of micro, small-, and medium-sized enterprises, including through access to financial services

8.4: Improve progressively, through 2030, global resource efficiency in consumption and production and endeavor to decouple economic growth from environmental degradation, in accordance with the 10-year framework of programs on sustainable consumption and production, with developed countries taking the lead

8.5: By 2030, achieve full and productive employment and decent work for all women and men, including for young people and persons with disabilities, and equal pay for work of equal value

8.6: By 2020, substantially reduce the proportion of youth not in employment, education, or training

8.7: Take immediate and effective measures to eradicate forced labor, end modern slavery and human trafficking, and secure the prohibition and elimination of the worst forms of child labor, including recruitment and use of child soldiers, and by 2025, end child labor in all its forms

8.8: Protect labor rights and promote safe and secure working environments for all workers, including migrant workers, in particular women migrants, and those in precarious employment

8.9: By 2030, devise and implement policies to promote sustainable tourism that creates jobs and promotes local culture and products

8.10: Strengthen the capacity of domestic financial institutions to encourage and expand access to banking, insurance, and financial services for all

8.a: Increase aid for trade support for developing countries, in particular least developed countries, including through the Enhanced Integrated Framework for Trade-Related Technical Assistance to Least Developed Countries

8.b: By 2020, develop and operationalize a global strategy for youth employment and implement the Global Jobs Pact of the International Labor Organization

Goal 9: Build Resilient Infrastructure, Promote Inclusive and Sustainable Industrialization, and Foster Innovation

Investment in sustainable infrastructure and in scientific and technological research increases economic growth, creates jobs, and promotes prosperity. Infrastructure projects costing billions are in the offing over the next 15 years, particularly in developing and emerging economies. Goal 9 therefore aims to build resilient infrastructure, promote industrialization, and foster innovation.

Increased resource-use efficiency and greater adoption of clean and environmentally sound technologies and industrial processes are necessary to make infrastructure and industries sustainable by 2030. Goal 9 aims to support technology development, research, and innovation especially in developing countries, provide small-scale industrial and other companies with greater access to financial services, including affordable credit, and increase the integration of these companies into value chains and markets. It also advocates providing universal and affordable access to the internet in the least developed countries of the world.

- 9.1: Develop quality, reliable, sustainable, and resilient infrastructure, including regional and transborder infrastructure, to support economic development and human well-being, with a focus on affordable and equitable access for all
- 9.2: Promote inclusive and sustainable industrialization and, by 2030, significantly raise the industry's share of employment and gross domestic product, in line with national circumstances, and double its share in least developed countries
- 9.3: Increase the access of small-scale industrial and other enterprises, in particular in developing countries, to financial services, including affordable credit, and their integration into value chains and markets
- 9.4: By 2030, upgrade infrastructure and retrofit industries to make them sustainable, with increased resource-use efficiency and greater adoption of clean and environmentally sound technologies and industrial processes, with all countries taking action in accordance with their respective capabilities
- 9.5: Enhance scientific research, upgrade the technological capabilities of industrial sectors in all countries, in particular developing countries, including encouraging innovation and substantially increasing the number of research and development workers per one million people and public and private research and development spending by 2030
- 9.a: Facilitate sustainable and resilient infrastructure development in developing countries through enhanced financial, technological, and technical support to African countries, least developed countries, landlocked developing countries, and small island developing states
- 9.b: Support domestic technology development, research, and innovation in developing countries, including by ensuring a conducive policy environment for, inter alia, industrial diversification and value addition to commodities
- 9.c: Significantly increase access to information and communications technology and strive to provide universal and affordable access to the internet in least developed countries by 2030

Goal 10: Reduce Inequality Within and Among Countries

Global inequalities are massive and present one of the biggest obstacles to sustainable development and the fight against poverty. Inequality within many countries has been rising in recent years. Inequalities limit the opportunities for social groups to participate in and make significant contributions to social, cultural, political, and economic life. Therefore, Goal 10 focuses on reducing inequality within and among countries.

In concrete terms, Goal 10 advocates delivering sustained income growth to the poorest 40% of the global population and achieving empowerment and social, economic, and political inclusion for all by 2030. Goal 10 aims to ensure equal opportunities through the elimination of discriminatory laws, policies, and practices, while facilitating orderly and safe human migration and mobility via the implementation of sound migration policies. It also envisages enhanced representation and a greater voice for developing countries in decision-making within international economic and financial institutions.

10.1: By 2030, progressively achieve and sustain income growth of the bottom 40 percent of the population at a rate higher than the national average

10.2: By 2030, empower and promote the social, economic, and political inclusion of all, irrespective of age, sex, disability, race, ethnicity, origin, religion, or economic or other status

10.3: Ensure equal opportunity and reduce inequalities of outcome, including by eliminating discriminatory laws, policies, and practices and promoting appropriate legislation, policies, and action in this regard

10.4: Adopt policies, especially fiscal, wage, and social protection policies, and progressively achieve greater equality

10.5: Improve the regulation and monitoring of global financial markets and institutions and strengthen the implementation of such regulations

10.6: Ensure enhanced representation and voice for developing countries in decision-making in global international economic and financial institutions in order to deliver more effective, credible, accountable, and legitimate institutions

10.7: Facilitate orderly, safe, regular, and responsible migration and mobility of people, including through the implementation of planned and well-managed migration policies

10.a: Implement the principle of special and differential treatment for developing countries, in particular least developed countries, in accordance with World Trade Organization agreements

10.b: Encourage official development assistance and financial flows, including foreign direct investment, to states where the need is greatest, in particular least developed countries, African countries, small island developing states, and landlocked developing countries, in accordance with their national plans and programs

10.c: By 2030, reduce to less than three percent the transaction costs of migrant remittances and eliminate remittance corridors with costs higher than five per cent

Goal 11: Make Cities and Human Settlements Inclusive, Safe, Resilient, and Sustainable

Urbanization is one of the most significant developments of the 21st century. More than half of the global population lives in cities, a proportion that is expected to increase to 70% by 2050. Cities drive local and national economies, as hubs of prosperity where more than 80% of global economic activity is concentrated. Urbanization also poses major challenges. Cities have an enormous environmental footprint. They occupy just three percent of the world's surface but are responsible for three-quarters of global resource consumption and 75% of global emissions.

Goal 11 aims to reduce the adverse per capita environmental impact of cities, particularly in terms of air quality and waste management. It calls for more inclusive and sustainable forms of urbanization, based in particular on a participatory, integrated, and sustainable approach to urban planning. In addition, it aims to ensure universal access to safe and inclusive green and public spaces, especially for women and children, older persons, and persons with disabilities, and provide access to safe and affordable housing and transport systems.

- 11.1: By 2030, ensure access for all to adequate, safe, and affordable housing and basic services and upgrade slums
- 11.2: By 2030, provide access to safe, affordable, accessible, and sustainable transport systems for all, improving road safety, notably by expanding public transport, with special attention to the needs of those in vulnerable situations, women, children, persons with disabilities, and older persons
- 11.3: By 2030, enhance inclusive and sustainable urbanization and capacity for participatory, integrated, and sustainable human settlement planning and management in all countries
- 11.4: Strengthen efforts to protect and safeguard the world's cultural and natural heritage
- 11.5: By 2030, significantly reduce the number of deaths and the number of people affected and substantially decrease the direct economic losses relative to global gross domestic product caused by disasters, including water-related disasters, with a focus on protecting the poor and people in vulnerable situations
- 11.6: By 2030, reduce the adverse per capita environmental impact of cities, including by paying special attention to air quality and municipal and other waste management
- 11.7: By 2030, provide universal access to safe, inclusive, and accessible, green and public spaces, in particular for women and children, older persons, and persons with disabilities
- 11.a: Support positive economic, social, and environmental links between urban, peri-urban, and rural areas by strengthening national and regional development planning
- 11.b: By 2020, substantially increase the number of cities and human settlements adopting and implementing integrated policies and plans toward inclusion, resource efficiency, mitigation and adaptation to climate change, resilience

to disasters, and develop and implement, in line with the Sendai Framework for Disaster Risk Reduction 2015–2030, holistic disaster risk management at all levels

11.c: Support least developed countries, including through financial and technical assistance, in building sustainable and resilient buildings utilizing local materials

Goal 12: Ensure Sustainable Consumption and Production Patterns

The world's population currently consumes more resources than ecosystems can provide. For social and economic development to remain within the carrying capacity of ecosystems, fundamental changes are necessary in the way societies produce and consume.

Goal 12 calls for the implementation of the UN's ten-year framework of programs on sustainable consumption and production patterns. It advocates environmentally sound management of chemicals and all waste and a substantial reduction in waste generation through measures such as recycling. Goal 12 also aims to halve food waste, encourage companies to adopt sustainable practices, and promote sustainable public procurement practices.

12.1: Implement the 10-Year Framework of programs on sustainable consumption and production patterns, all countries taking action, with developed countries taking the lead, taking into account the development and capabilities of developing countries

12.2: By 2030, achieve the sustainable management and efficient use of natural resources

12.3: By 2030, halve per capita global food waste at the retail and consumer levels and reduce food losses along production and supply chains, including post-harvest losses

12.4: By 2020, achieve the environmentally sound management of chemicals and all wastes throughout their life cycle, in accordance with agreed international frameworks, and significantly reduce their release to air, water, and soil in order to minimize their adverse impacts on human health and the environment

12.5: By 2030, substantially reduce waste generation through prevention, reduction, recycling, and reuse

12.6: Encourage companies, especially large and transnational companies, to adopt sustainable practices and to integrate sustainability information into their reporting cycle

12.7: Promote public procurement practices that are sustainable, in accordance with national policies and priorities

12.8: By 2030, ensure that people everywhere have the relevant information and awareness for sustainable development and lifestyles in harmony with nature

12.a: Support developing countries to strengthen their scientific and technological capacity to move toward more sustainable patterns of consumption and production

12.b: Develop and implement tools to monitor sustainable development impacts for sustainable tourism that creates jobs and promotes local culture and products

12.c: Rationalize inefficient fossil-fuel subsidies that encourage wasteful consumption by removing market distortions, in accordance with national circumstances, including by restructuring taxation and phasing out those harmful subsidies, where they exist, to reflect their environmental impacts, taking fully into account the specific needs and conditions of developing countries and minimizing the possible adverse impacts on their development in a manner that protects the poor and the affected communities

Goal 13: Take Urgent Action to Combat Climate Change and Its Impacts

Climate change is a key challenge with regard to sustainable development. The warming of the earth's atmosphere is triggering changes in the global climate system that threaten the livelihoods of large sections of the population in less developed countries, while infrastructure and certain economic sectors in particular are vulnerable to the risks of climate change in developed regions. Furthermore, changes in precipitation and temperature cycles are also affecting ecosystems such as forests, agricultural land, mountain regions, and oceans, as well as the plants, animals, and people that live in them. Global carbon dioxide (CO_2) emissions increased by over 50% between 1990 and 2012.

Goal 13 calls on countries to incorporate climate protection measures in their national policies and assist each other in responding to the challenges at hand. It acknowledges that the United Nations Framework Convention on Climate Change is the primary international, intergovernmental forum for negotiating the global response to climate change. Supplementing this dialogue, Goal 13 advocates strengthening resilience to climate-related natural disasters and reaffirms the commitment undertaken by developed countries to mobilize each year USD 100 billion jointly from all sources by 2020 to help developing countries adapt to climate change.

13.1: Strengthen resilience and adaptive capacity to climate-related hazards and natural disasters in all countries

13.2: Integrate climate change measures into national policies, strategies, and planning

13.3: Improve education, awareness-raising, and human and institutional capacity on climate change mitigation, adaptation, impact reduction, and early warning

13.a: Implement the commitment undertaken by developed-country parties to the United Nations Framework Convention on Climate Change to a goal of mobilizing jointly $100 billion annually by 2020 from all sources to address the needs of developing countries in the context of meaningful mitigation actions and transparency on implementation and fully operationalize the Green Climate Fund through its capitalization as soon as possible

13.b: Promote mechanisms for raising capacity for effective climate change-related planning and management in least developed countries and small island developing states, including focusing on women, youth, and local and marginalized communities

Goal 14: Conserve and Sustainably Use the Oceans, Seas, and Marine Resources for Sustainable Development

Pollution and over-exploitation of our oceans are posing ever-greater problems, such as an acute threat to biodiversity, ocean acidification, and an increase in plastic waste. Besides industrial fishing and the commercial use of marine resources, climate change is placing marine ecosystems under increased pressure. A continuously growing global population will be even more dependent on marine resources in the future.

Goal 14 advocates significantly reducing all kinds of marine pollution and minimizing ocean acidification by 2025, as well as sustainably managing and protecting marine and coastal ecosystems by as early as 2020. It also aims, by 2020, to regulate harvesting in an effective manner and to halt overfishing by ending illegal and unregulated fishing and destructive fishing practices. In addition, Goal 14 aims to prohibit specific types of subsidies to fisheries.

14.1: By 2025, prevent and significantly reduce marine pollution of all kinds, in particular from land-based activities, including marine debris and nutrient pollution

14.2: By 2020, sustainably manage and protect marine and coastal ecosystems to avoid significant adverse impacts, including by strengthening their resilience, and take action for their restoration in order to achieve healthy and productive oceans

14.3: Minimize and address the impacts of ocean acidification, including through enhanced scientific cooperation at all levels

14.4: By 2020, effectively regulate harvesting and end overfishing; illegal, unreported, and unregulated fishing; and destructive fishing practices and implement science-based management plans in order to restore fish stocks in the shortest time feasible, at least to levels that can produce maximum sustainable yield as determined by their biological characteristics

14.5: By 2020, conserve at least ten percent of coastal and marine areas, consistent with national and international law and based on the best available scientific information

14.6: By 2020, prohibit certain forms of fisheries subsidies which contribute to overcapacity and overfishing; eliminate subsidies that contribute to illegal, unreported, and unregulated fishing; and refrain from introducing new such subsidies, recognizing that appropriate and effective special and differential treatment for developing and least developed countries should be an integral part of the World Trade Organization fisheries subsidies negotiation

14.7: By 2030, increase the economic benefits to small island developing states and least developed countries from the sustainable use of marine resources, including through sustainable management of fisheries, aquaculture, and tourism

14.a: Increase scientific knowledge, develop research capacity, and transfer marine technology, taking into account the Intergovernmental Oceanographic Commission Criteria and Guidelines on the Transfer of Marine Technology, in order to improve ocean health and to enhance the contribution of marine biodiversity to the development of developing countries, in particular small island developing states and least developed countries

14.b: Provide access for small-scale artisanal fishers to marine resources and markets

14.c: Enhance the conservation and sustainable use of oceans and their resources by implementing international law as reflected in the United Nations Convention on the Law of the Sea, which provides the legal framework for the conservation and sustainable use of oceans and their resources, as recalled in paragraph 158 of "The future we want"

Goal 15: Protect, Restore, and Promote Sustainable Use of Terrestrial Ecosystems, Sustainably Manage Forests, Combat Desertification, and Halt and Reverse Land Degradation and Halt Biodiversity Loss

The conservation and sustainable use of biodiversity are vital to social and economic development as well as to humanity's survival. However, there is evidence of an ongoing decline in biodiversity along with a loss in forest area that threatens human prosperity, with poor rural populations – including indigenous and local communities – particularly affected. Biodiversity and forests contribute to poverty reduction, for example by underpinning food security and human health, providing clean air and water, absorbing CO_2 emissions, and providing a basis for environmental development.

Goal 15 calls for the conservation, restoration, and sustainable use of ecosystems, with the aim of halting deforestation, restoring degraded forests, and substantially increasing reforestation by 2020. It also advocates combating desertification by 2030 and restoring land affected by desertification, drought, and floods. To protect biodiversity, Goal 15 calls for urgent measures to end poaching and trafficking of protected plant and animal species.

15.1: By 2020, ensure the conservation, restoration, and sustainable use of terrestrial and inland freshwater ecosystems and their services, in particular forests, wetlands, mountains, and drylands, in line with obligations under international agreements

15.2: By 2020, promote the implementation of sustainable management of all types of forests, halt deforestation, restore degraded forests, and substantially increase afforestation and reforestation globally

15.3: By 2030, combat desertification, restore degraded land and soil, including land affected by desertification, drought, and floods, and strive to achieve a land degradation-neutral world

15.4: By 2030, ensure the conservation of mountain ecosystems, including their biodiversity, in order to enhance their capacity to provide benefits that are essential for sustainable development

15.5: Take urgent and significant actions to reduce the degradation of natural habitats, halt the loss of biodiversity, and, by 2020, protect and prevent the extinction of threatened species

15.6: Promote fair and equitable sharing of the benefits arising from the utilization of genetic resources and promote appropriate access to such resources, as internationally agreed
15.7: Take urgent action to end poaching and trafficking of protected species of flora and fauna and address both demand and supply of illegal wildlife products
15.8: By 2020, introduce measures to prevent the introduction and significantly reduce the impact of invasive alien species on land and water ecosystems and control or eradicate the priority species
15.9: By 2020, integrate ecosystem and biodiversity values into national and local planning, development processes, poverty reduction strategies, and accounts
15.a: Mobilize and significantly increase financial resources from all sources to conserve and sustainably use biodiversity and ecosystems
15.b: Mobilize significant resources from all sources and at all levels to finance sustainable forest management and provide adequate incentives to developing countries to advance such management, including for conservation and reforestation
15.c: Enhance global support for efforts to combat poaching and trafficking of protected species, including by increasing the capacity of local communities to pursue sustainable livelihood opportunities

Goal 16: Promote Peaceful and Inclusive Societies for Sustainable Development; Provide Access to Justice for All; and Build Effective, Accountable and Inclusive Institutions at All Levels

Evidence shows that without peaceful and inclusive communities and effective governance, development cannot be sustainable. For example, conflict-affected countries are the furthest away from achieving the MDGs, whereas in many other countries, the re-establishment of peace and accountable institutions has contributed greatly to achieving the MDGs.

Goal 16 for 2030 therefore aims to promote peaceful and inclusive societies. As such, it advocates reducing all forms of violence, ending torture, and combating all forms of organized crime. In addition, Goal 16 envisages significantly reducing corruption and bribery as well as illicit financial and arms flows. To ensure that societies are peaceful and inclusive, Goal 16 also aims to promote inclusive institutions and the rule of law and guarantee equal access to justice.

16.1: Significantly reduce all forms of violence and related death rates everywhere
16.1: End abuse, exploitation, trafficking, and all forms of violence against and torture of children
16.3: Promote the rule of law at the national and international levels and ensure equal access to justice for all
16.4: By 2030, significantly reduce illicit financial and arms flows, strengthen the recovery and return of stolen assets, and combat all forms of organized crime

16.5: Substantially reduce corruption and bribery in all their forms
16.6: Develop effective, accountable, and transparent institutions at all levels
16.7: Ensure responsive, inclusive, participatory, and representative decision-making at all levels
16.8: Broaden and strengthen the participation of developing countries in the institutions of global governance
16.9: By 2030, provide legal identity for all, including birth registration
16.10: Ensure public access to information and protect fundamental freedoms, in accordance with national legislation and international agreements
16.a: Strengthen relevant national institutions, including through international cooperation, for building capacity at all levels, in particular in developing countries, to prevent violence and combat terrorism and crime
16.b: Promote and enforce non-discriminatory laws and policies for sustainable development

Goal 17: Strengthen the Means of Implementation and Revitalize the Global Partnership for Sustainable Development

Successful implementation of the 17 SDGs depends on a comprehensive funding framework that goes beyond official development assistance commitments. Alongside public and private funding, the political sphere will also be expected to make a greater contribution to achieving the goals in question. In July 2015, the international community agreed to a new framework for financing and implementing sustainable development – the Addis Ababa Action Agenda.

Goal 17 calls on developed countries to renew their commitment to allocate 0.7% of their gross national income to official development assistance. It aims for a greater mobilization of domestic resources to reduce dependence on foreign support, as well as enhanced international collaboration in science, technology, and innovation, and the promotion of an equitable multilateral trading system. Goal 17 also advocates enhancing macroeconomic stability and policy coherence in the interests of sustainable development.

Finance

- Strengthen domestic resource mobilization, including through international support to developing countries, to improve domestic capacity for tax and other revenue collection
- Developed countries to implement fully their official development assistance commitments, including the commitment by many developed countries to achieve the target of 0.7 percent of gross national income for official development assistance (ODA/GNI) to developing countries and 0.15–0.20 percent of ODA/GNI to least developed countries; ODA providers are encouraged to consider setting a target to provide at least 0.20 percent of ODA/GNI to least developed countries
- Mobilize additional financial resources for developing countries from multiple sources

- Assist developing countries in attaining long-term debt sustainability through coordinated policies aimed at fostering debt financing, debt relief, and debt restructuring, as appropriate, and address the external debt of highly indebted poor countries to reduce debt distress
- Adopt and implement investment promotion regimes for least developed countries

Technology

- Enhance North-South, South-South, and triangular regional and international cooperation on and access to science, technology, and innovation and enhance knowledge sharing on mutually agreed terms, including through improved coordination among existing mechanisms, in particular at the UN level, and through a global technology facilitation mechanism
- Promote the development, transfer, dissemination, and diffusion of environmentally sound technologies to developing countries on favorable terms, including on concessional and preferential terms, as mutually agreed
- Fully operationalize the technology bank and science, technology, and innovation capacity-building mechanism for least developed countries by 2017 and enhance the use of enabling technology, in particular information and communications technology

Capacity-building

- Enhance international support for implementing effective and targeted capacity-building in developing countries to support national plans to implement all the SDGs, including through North-South, South-South, and triangular cooperation

Trade

- Promote a universal, rules-based, open, non-discriminatory, and equitable multilateral trading system under the World Trade Organization, including through the conclusion of negotiations under its Doha Development Agenda
- Significantly increase the exports of developing countries, in particular with a view to doubling the least developed countries' share of global exports by 2020
- Realize timely implementation of duty-free and quota-free market access on a lasting basis for all least developed countries, consistent with World Trade Organization decisions, including by ensuring that preferential rules of origin applicable to imports from least developed countries are transparent and simple, and contribute to facilitating market access

Systemic issues

Policy and institutional coherence

- Enhance global macroeconomic stability, including through policy coordination and policy coherence
- Enhance policy coherence for sustainable development

- Respect each country's policy space and leadership to establish and implement policies for poverty eradication and sustainable development

Multi-stakeholder partnerships

- Enhance the global partnership for sustainable development, complemented by multi-stakeholder partnerships that mobilize and share knowledge, expertise, technology, and financial resources, to support the achievement of the SDGs in all countries, in particular developing countries
- Encourage and promote effective public, public-private, and civil society partnerships, building on the experience and resourcing strategies of partnerships

Data, monitoring, and accountability

- By 2020, enhance capacity-building support to developing countries, including for least developed countries and small island developing states, to significantly increase the availability of high-quality, timely, and reliable data disaggregated by income, gender, age, race, ethnicity, migratory status, disability, geographic location, and other characteristics relevant in national contexts
- By 2030, build on existing initiatives to develop measurements of progress on sustainable development that complement gross domestic product and support statistical capacity-building in developing countries

NAE's 14 grand challenges

Along a similar line to the creation of the UN's SDGs, in 2008, the USA Academy of Engineering presented the 14 Grand Challenges for engineering in combating world problems. The narratives of the 14 grand challenges are summarized below.

1. <u>Make solar energy economical</u>: Solar energy provides less than one percent of the world's total energy, but it has the potential to provide much, much more.
2. <u>Provide energy from fusion</u>: Human-engineered fusion has been demonstrated on a small scale. The challenge is to scale up the process to commercial proportions, in an efficient, economical, and environmentally benign way.
3. <u>Develop carbon sequestration methods</u>: Engineers are working on ways to capture and store excess CO_2 to prevent global warming.
4. <u>Manage the nitrogen cycle</u>: Engineers can help restore balance to the nitrogen cycle with better fertilization technologies and by capturing and recycling waste.
5. <u>Provide access to clean water</u>: The world's water supplies are facing new threats; affordable, advanced technologies could make a difference for millions of people around the world.
6. <u>Restore and improve urban infrastructure</u>: Good design and advanced materials can improve transportation and energy, water, and waste systems and also create more sustainable urban environments.

7. Advance health informatics: Stronger health information systems not only improve everyday medical visits, but they are essential to counter pandemics and biological or chemical attacks.
8. Engineer better medicines: Engineers are developing new systems to use genetic information, sense small changes in the body, assess new drugs, and deliver vaccines.
9. Reverse-engineer the brain: The intersection of engineering and neuroscience promises great advances in health care, manufacturing, and communication.
10. Prevent nuclear terror: The need for technologies to prevent and respond to a nuclear attack is growing.
11. Secure cyberspace: It's more than preventing identity theft. Critical systems in banking, national security, and physical infrastructure may be at risk.
12. Enhance virtual reality: True virtual reality creates the illusion of actually being in a different space. It can be used for training, treatment, and communication.
13. Advance personalized learning: Instruction can be individualized based on learning styles, speeds, and interests to make learning more reliable.
14. Engineer the tools of scientific discovery: In the century ahead, engineers will continue to be partners with scientists in the great quest for understanding many unanswered questions of nature.

APPLICABILITY OF INDUSTRIAL ENGINEERING

As can be seen in the narratives of the SDGs, particularly Goal 17, all the ideals of industrial engineering are engendered in the desired implementation pathways. A similar argument exists for the 14 Grand Challenges articulated by the NAE. Thus, we have a double alignment basis for industrial engineering. At this point, it is instructive to present condensed definitions of industrial and systems engineering.

> **Industrial engineering** is concerned with the design, installation, and improvement of integrated systems of people, materials, information, equipment, and energy by drawing upon specialized knowledge and skills in the mathematical, physical, and social sciences, together with the principles and methods of engineering analysis and design to specify, predict, and evaluate the results to be obtained from such systems.

The human element is very important to industrial engineers. They deal with people from all levels in an organization. Recalling industrial engineers' pursuit of improvement of systems, a common slogan is that ***"engineers make things, but industrial engineers make things better."*** Based on this, the tools and techniques of industrial engineering, prefaced by technology, are essential for accomplishing the elements of the UN's SDGs. The goal relating to capacity-building is, particularly, within the realm of industrial engineering.

> A system is a collection of interrelated elements, whose collective output, produced in unison, is higher than the mere sum of the individual outputs of the elements.

In other words, a group of people working together can achieve more than a serial linking of individual achievements. Industrial engineers operate under the premise of this definition of a system.

> Systems engineering is the application of engineering to solutions of a multi-faceted problem through a systematic collection and integration of parts of the problem with respect to the lifecycle of the problem. It is the branch of engineering concerned with the development, implementation, and use of large or complex systems. It focuses on specific goals of a system considering the specifications, prevailing constraints, expected services, possible behaviors, and structure of the system. It also involves a consideration of the activities required to assure that the system's performance matches the stated goals. Systems engineering addresses the integration of tools, people, and processes required to achieve a cost-effective and timely operation of the system.

Taken together, the above concepts and definitions are important to achieving and measuring progress with the 17 SDGs. All engineers make things. Industrial engineers not only make things, they make things better through better efficiency, effectiveness, quality, performance, and productivity. The platform of enhancements provided by industrial engineering includes products, services, and results. All the desirable dimensions of performance are embodied within the practice of industrial engineering. With the systems-based approach of industrial engineering, SDGs can be executed faster, efficiently, effectively, economically, and safely too.

Based on the advent of AI applications, we expect human-centered technologies to plan more active roles in the pursuit of the UN's SDGs. Industrial engineering is poised to be at the forefront of such new strategies. Industrial and cognitive revolutions will make SDGs more achievable through the combination and permutation of human-to-human, human-to-AI, robot-to-robot, and robot-to-human teaming strategies. Our strategic recommendations include the following.

- Create more public-private partnerships
- Invest more in autonomous systems
- Get academia more involved in policy development and implementation
- Enmesh science and technology more seamlessly
- Increase science and technology transfer pathways
- Coordinate managerial processes to mitigate risk
- Embrace and leverage quantum computing systems
- Recognize and mitigate climate change impacts
- Modernize and technologize workforce development
- Invest in government-industry-military-business-academia collaboration
- Recapitalize infrastructure
- Put societal needs within business strategies

From an industrial engineering standpoint, whatever affects one subsystem of the global infrastructure will eventually percolate through the whole system. In 2008, the NAE published the "14 Grand Challenges for Engineering." The challenges have global implications for everyone, not just the engineering professions. As such, solution strategies must embrace all disciplines. Industrial engineering, by virtue of its versatility and systems viewpoint, can be the anchor for solving the challenges. An extract from the NAE document on the 14 grand challenges reads:

> In sum, governmental and institutional, political and economic, and personal and social barriers will repeatedly arise to impede the pursuit of solutions to problems. As they have throughout history, engineers will have to integrate their methods and solutions with the goals and desires of all society's members.

This statement emphasizes the relevance of a holistic systems-thinking approach in solving the multi-faceted global problems that we face now and will face in the future. This is in perfect alignment with the definition and practice of industrial engineering, as embodied in the definition presented earlier. The approach of industrial engineering, using a systems perspective, is how we move from rhetoric to action. The definition of industrial engineering embodies what is needed to develop and implement solutions to the 14 grand challenges, which involve design, installation, improvement, systems integration, people, materials, information, energy, prediction, evaluation, and implementation.

A semantic network is a good way to represent the systems interrelationships of the NAE's 14 grand challenges. We offer Figure 1.2 as an illustrative example of how

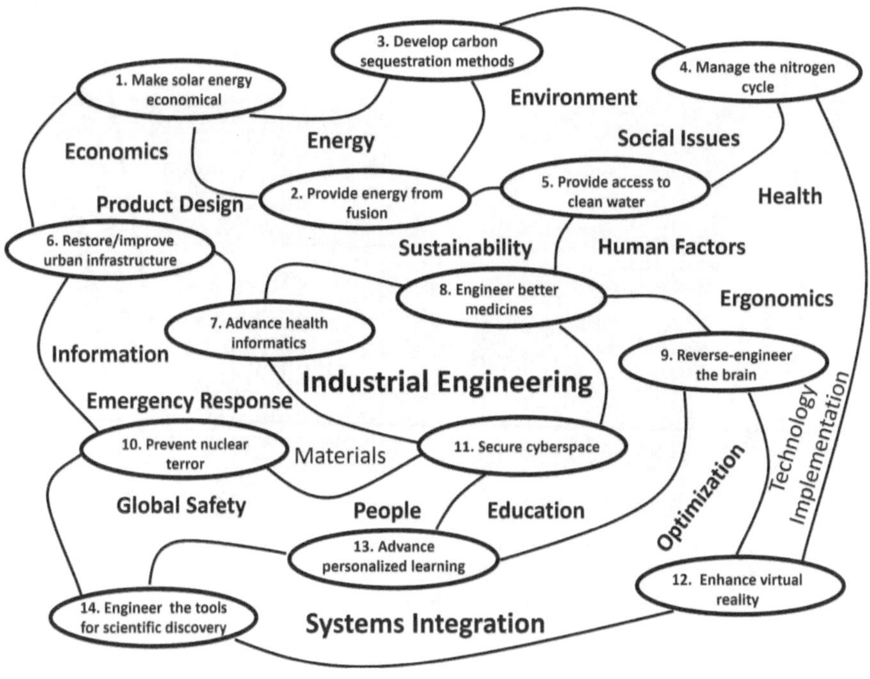

FIGURE 1.2 A Semantic Network Representation of the NAE's 14 Grand Challenges.

every element of the challenges is linked to at least two other elements with the diverse practice of industrial engineering. The overall problem sphere is composed of a system of systems (SoS) structure. Using the tools of engineering requirements analysis, industrial engineers effectively ferret out underlying relationships among the elements of a problem, thereby ensuring that all factors are considered in a global holistic solution.

SYSTEMS STRUCTURE

The V-model of systems engineering is used by industrial engineers to link and solve complex problems. It is a graphical representation of a system's lifecycle. The model, which relates solution techniques to problem scenarios, is directly applicable to the semantic network of the grand challenges. Figure 1.3 presents a V-model in the context of solving the grand challenges. As indicated in the figure, problem assessment, problem decomposition, lifecycle requirements, solution integration, and global implementation are key elements of solving the 14 grand challenges (see Badiru 2013a, Badiru 2013b, Badiru et al 2024a, and Badiru et al 2024b).

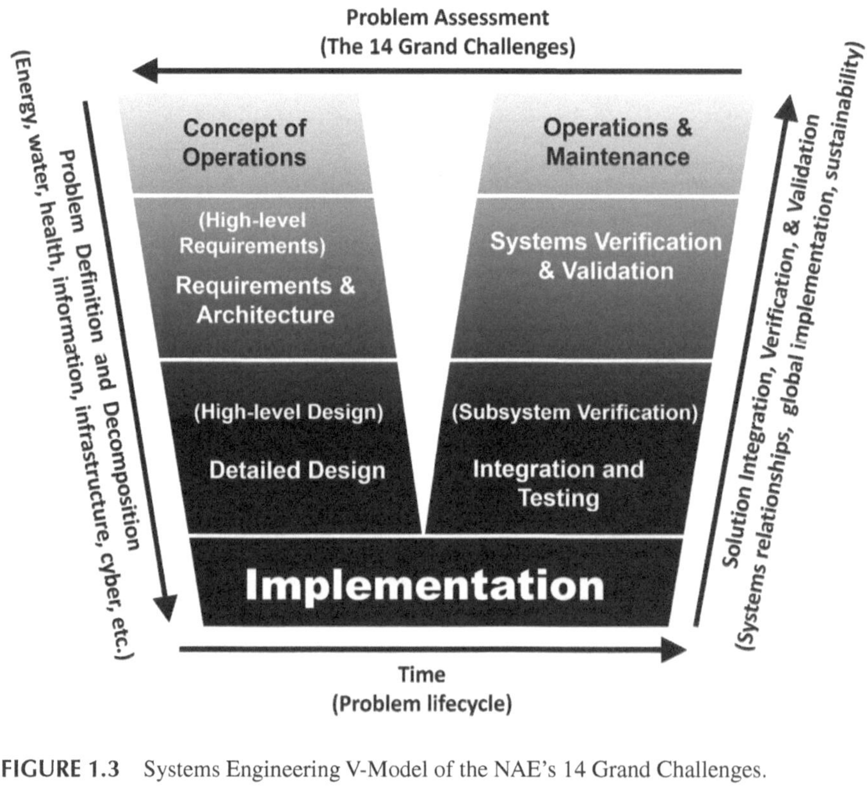

FIGURE 1.3 Systems Engineering V-Model of the NAE's 14 Grand Challenges.

THE COMMON THEME OF ENERGY

Several case examples of industrial engineering applications exist for the various elements and stages of the 14 grand challenges. For the purpose of an illustrative example, we identify the specific case of energy, which appears to be a common theme in many of the challenges of the world. The following are a few examples of what industrial engineers are doing in addressing global energy issues.

- Industrial engineers use the concept of SoS to define functional, physical, and behavioral requirements in designing energy management plans for business and industry. The industrial engineering team, in the Systems Engineering and Management department at the AFIT, collaborates with other disciplines to develop energy management and conservation strategies for jet fuel price hedging for USA air force fighter fleet.
- During the energy crisis of the 1970s, industrial engineers were called upon to develop company strategies to ensure business survival at the height of the crisis. There is the case of Westmoreland Coal Company hiring an industrial engineer, Dr. Sid Gilbreath, as a General Manager in Virginia to develop and oversee their energy business operations in the early 1980s.
- Industrial engineering researchers at Oregon State University work collaboratively and in interdisciplinary teams to create and deploy technologies that generate, distribute, store, consume, and recycle energy in a clean and efficient manner. They address energy and sustainability challenges in developing systems and products that meet the needs of society by considering societal, environmental, and economic requirements. Areas of interdisciplinary research involving industrial engineering include marine and wind renewable energy, waste heat recovery, combustion energy conversion, and novel transportation fuels, high heat flux thermal management in micro-scale systems, spray cooling using nano-fluids and nanostructures, two-phase flow for thermal control, multilayer thin film structures for high energy density micro-capacitors, refractory metal-based alloys for efficient power generation, micro-reactor-assisted nanomaterial deposition for photovoltaic thin films, process design for recycling hospital waste streams, sustainability dimensions of human work, environmentally sustainable nano-materials and nano-manufacturing, micro-channel-based solar biofuel processing, and sustainable fiber composites for energy technology.
- The Smart Systems and Operations Laboratory with the School of Industrial Engineering at Purdue University is involved in research activities in design, modeling, monitoring, and control of complex systems of systems and their operations, with applications to energy systems in diverse operations. Using a systems structure, the lab's research investigates complex design, integration, monitoring, and control of systems.
- A team of industrial engineering researchers, in collaboration with the Sarkeys Energy Center at the University of Oklahoma, Norman, is

addressing a hybrid systems engineering approach to study the technical, managerial, physical, and economic aspects of integrated energy systems.

- In addition to technical issues, industrial engineers are already involved in product design and policy issues related to energy challenges. The *Industrial Engineer* magazine (Badiru, 2008b) published guidelines for energy requirements analysis for new product development in response to energy challenges of the following.
  - **Generation**: Fossil, fission, fusion, renewable, safety, security, and energy independence
  - **Distribution**: Transmission technology, hydrogen, distributed energy sources, and market
  - **Consumption**: Transportation, storage, product requirements, and conservation

These are just a few examples of how industrial engineers are already playing direct roles and developing solutions that are in line with the requirements of the NAE's 14 grand challenges. Similar examples exist for all facets of issues contained in the NAE's statements and the UN's SDGs. Technology transfer, economics of design, big data management, additive manufacturing, health systems engineering, environmental impact assessment, solar economics, water resources, cyber infrastructure, and learning systems are among the other areas where industrial engineering is already established as a solution source. What is important now is to develop a compendium of existing industrial engineering body of work that can be leveraged collaboratively to position the discipline at the forefront of global solution strategies. The premise of this book fits that goal. Table 1.1 presents a Taxonomy of Industrial Engineering (IE)'s alignment with the grand challenges. The entries in the table are not mutually exclusive or exhaustive. Researchers, educators, and practitioners may expand the table to fit specific ongoing activities. It would be helpful to conduct further research on the table to add a column of who and who (IE groups or others) have active ongoing work in each area of the grand challenges, as well as the UN SDGs.

IMPLEMENTATION RECOMMENDATIONS

Engineers of the future will need diverse skills to tackle the multitude of issues and factors involved in adequately and successfully addressing the 14 grand challenges. In January 2013, Adedeji Badiru served on a panel at the World Congress on Engineering Education (2013a) in Doha, Qatar. His presentation was on "Fourteen Grand Challenges and 2020 Skills for Engineers," in which he discussed the critical skills that engineers must have in the 2020 timeframe in order to contribute to devising sustainable solutions for global problems, problems that affect everyone in every corner of the world. Industrial engineering skills and knowledge of project execution techniques will be the cornerstone for integrating multiple skills across geographical boundaries as well as across cultural divides. Using a socio-technical systems framework, industrial engineers can facilitate new products, services, and results efficiently within cost and schedule constraints to address the 14 grand

TABLE 1.1
Taxonomical Analysis of Industrial Engineering's Alignment with The NAE's Grand Challenges

Grand Challenge	IE Research Area	IE Education Area	IE Practice Area
1. Make solar energy economical	Strategic investments	Engineering economics	Portfolio management
2. Provide energy from fusion	Design safety	Energy management	Energy strategies
3. Develop carbon sequestration methods	Natural Science analytics	Storage systems design	Global systems interfaces
4. Manage the nitrogen cycle	System planning	Systems optimization	Environmental science
5. Provide access to clean water	Environmental science	Water resource engineering	Water management
6. Restore and improve urban infrastructure	Resilience engineering	Construction management	Infrastructure design
7. Advance health informatics	Health systems engineering	Bio informatics	Industrial hygiene
8. Engineer better medicines	Biomedical engineering	Product quality control	Occupational health
9. Reverse-engineer the brain	Human factors	Cognitive psychology	Behavioral science
10. Prevent nuclear terror	Deterrent strategies	Emergency response	Social responsibility and negotiation
11. Secure cyberspace	Information resources research	Enterprise design and management	Information security
12. Enhance virtual reality	Software engineering	Software design and programming	Game design
13. Advance personalized learning	Active learner systems	Blended hybrid learning	Distance learning systems
14. Engineer the tools of scientific discovery	Product design	Scientific inquiry	Ergonomics

challenges and the UN's SDGs. Susan Blake, an industrial engineering colleague, surmises that:

> IEs make systems function better together with less waste, better quality, and fewer resources.

This is exactly what the 14 grand challenges need. The goal of every organization is to eliminate waste in the pursuit of objectives. The challenges should not be approached from the perspective of technical silos alone, as presently practiced. They must be addressed from the perspective of holistic systems. Industrial engineering educators and practitioners must come together to develop robust curricula to adequately prepare students for the engineering challenges that lie ahead. Together, we can accomplish the ideals postulated by the NAE in compiling the 14 grand challenges. It has been several years since the 14 grand challenges were introduced, yet sustainable accomplishments are still on the horizon, perhaps even a mirage. A sweeping collaborative systems approach should not wait any longer. A summary of recommendations are:

- IEs need to get more involved in the science, engineering, and policy issues at the national level. The platform of sustainable results can be built only on sound public policy to actualize what scientists and engineers envision.
- IE groups need to band together to present a cohesive and collective front to contribute to solutions to the grand challenges.
- What IEs are already doing needs to be articulated and promoted uniformly at the national level.
- IE researchers need to proactively volunteer and/or initiate collaborative alliances across disciplines in interdisciplinary efforts, whereby everyone brings something to the table.
- Everyone involved should keep the systems view within view at all times because everything is interconnected.

The various issues, strategies, and summations presented in this chapter have been addressed over the years on different platforms, including Maxwell (2024), Badiru et al. (2024a), Badiru and Tourangeau (2024), Badiru (2024a, 2024b), Peacock and Badiru (2023), Badiru and Asaolu (2023), Badiru and Tourangeau (2023), Badiru and Omitaomu (2023), Badiru (2023a, 2023b, 2023c), Badiru and Mertens (2023), Doran et al. (2022), Badiru (2022a, 2022b, 2022c, 2022d), Badiru and Lamont (2022), Badiru and Cromarty (2021), Badiru (2021a, 2021b, 2021c, 2021d), Badiru and Agustiady (2021), Badiru (2020), Badiru (2015), Badiru (2014), Badiru (2013b), and Badiru (2008a, 2008b). Interested readers may consult these references for further details as may be desired.

REFERENCES

Badiru, A. B. (2008b), *Triple C Model of Project Management: Communication, Cooperation, and Coordination*, Taylor & Francis Group/CRC Press, Boca Raton, FL.

Badiru, Adedeji B. (2008a), "New Power Generation: Energy Requirements Planning Must Drive Future Product Development," *Industrial Engineer*, September 2008, pp. 36–40.

Badiru, A. B. (2013a), "*NAE's 14 Grand Challenges and 2020 Skills for Engineers*," *Panel Discussion Keynote, World Congress on Engineering Education (WCEE2013)*, Doha, Qatar, January 7–9.

Badiru, Adedeji B. (2013b), "Up to the Task: Industrial engineering is essential to meeting NAE's grand challenges," *Industrial Engineer*, Vol. 45, No. 11, pp. 42–45, November 2013.

Badiru, Adedeji B. (2014), "Beyond Swords and Plowshares: Military service can advance engineering education and address grand challenges," *ASEE PRISM*, October 2014, p. 64.

Badiru, Adedeji B. (2015), "A Recipe for Meeting Broader Global Societal Challenges: STEM Education for the Service member has benefits now and in the future," *Military Advanced Education & Transition*," September 2015, Vol. 10, Issue 7/8, pp. 15–17.

Badiru, Adedeji B. (2020), *Innovation: A Systems Approach*, Taylor & Francis Group/CRC Press, Boca Raton, FL.

Badiru, Adedeji B. (2021a), *Project Management Essentials: Analytics for Control*, Taylor & Francis Group/CRC Press, Boca Raton, FL.

Badiru, Adedeji B. (2021b), "The Flight of the Billionaires: Collateral Engineering Benefits of Commercial Space Flight," *ASEE PRISM*, October 2021, p. 44.

Badiru, Adedeji B. (2021c), "Revitalizing Workforce Development in Education," in *SOCHE News* Online (Strategic Ohio Council for Higher Education), 5 April 2021.

Badiru, Adedeji B. (2021d), "World Systems View of the Environment: The Essential Role of the United States Military," *The AFIT Engineer*, September 2021, pp. 1–2.

Badiru, Adedeji B. (2022a), *Global Supply Chain: Using Systems Engineering Strategies to Respond to Disruptions*, Taylor & Francis Group/CRC Press, Boca Raton, FL.

Badiru, Adedeji B. (2022b), *Artificial Intelligence and Digital Systems Engineering*, Taylor & Francis Group/CRC Press, Boca Raton, FL.

Badiru, Adedeji B. (2022c), "The next frontier for Industrial Engineering," *ISE Magazine*, September 2022, p. 24.

Badiru, Adedeji B. (2022d), "Digital Transformation: Universities Often First Adopters of Latest Technological Advancements," *The AFIT Engineer*, Vol. 4, Issue 2, June 2022, pp. 1–2.

Badiru, Adedeji B. (2023a), "Transforming IE through Academic Leadership," *ISE Magazine*, August 2023, p. 26.

Badiru, Adedeji B. (2023b), "Industrial Engineering Hope for Nigeria," *ISE Magazine*, March 2023, p. 26.

Badiru, Adedeji B. (2023c), *Systems Engineering Using DEJI Systems Model: Design, Evaluation, Justification, and Integration with Case Studies and Applications*, Taylor & Francis Group/CRC Press, Boca Raton, FL.

Badiru, Adedeji B. (2024a), *Systems Engineering: Influencing Our Planet and Re-Engineering Our Actions*, Taylor & Francis Group/CRC Press, Boca Raton, FL.

Badiru, Adedeji B. (2024b), "CHIP on our shoulder requires strategy for workforce development," *ISE Magazine*, June 2024, p. 24.

Badiru, Adedeji B. and Tina Agustiady (2021), *Sustainability: A Systems Engineering Approach to the Global Grand Challenge*, Taylor & Francis Group/CRC Press, Boca Raton, FL.

Badiru, Adedeji B. and Olumuyiwa S. Asaolu (2023), *Handbook of Mathematical and Digital Engineering Foundations for Artificial Intelligence: A Systems Methodology*, Taylor & Francis Group/CRC Press, Boca Raton, FL.

Badiru, Adedeji B., Frank Ciarallo, and Eric Mbonimpa (2023), *Handbook of Scholarly Publications from the Air Force Institute of Technology (AFIT)*, Taylor & Francis Group/CRC Press, Boca Raton, FL.

Badiru, Adedeji B. and Lauralee Cromarty (2021), *Operational Excellence in the New Digital Era*, Taylor & Francis Group/CRC Press, Boca Raton, FL.

Badiru, Adedeji B. and Gary Lamont (2022), *Innovation Fundamentals: Quantitative and Qualitative Techniques*, Taylor & Francis Group/CRC Press, Boca Raton, FL.

Badiru, Adedeji B. and Andreas Mertens (2023), "TELEWORK: Possible Learning Curve Effects and the Need for More Analysis," *Defense Acquisition Magazine*, March-April, 2023, pp. 14–18.

Badiru, Adedeji B. and O. A. Omitaomu (2023), *Systems 4.0: System Foundations for Industry 4.0*, Taylor & Francis Group/CRC Press, Boca Raton, FL.

Badiru, Adedeji B. and Melinda L. Tourangeau (2023), *Leadership Matters: Industrial Engineering Framework for Developing and Sustaining Industry*, Taylor & Francis Group/CRC Press, Boca Raton, FL.

Badiru, Adedeji B. and Melinda L. Tourangeau (2024), *Assessing Innovation: Metrics, Rubrics, and Standards*, Taylor & Francis Group/CRC Press, Boca Raton, FL

Badiru, Adedeji, Wagenknecht, Nils, Andreas Mertens, and Olufemi Omitaomu (2024a), "Legal and Business Aspects of Climate Agreements," Book Chapter 6 in Badiru, Adedeji B. (2024), Systems Engineering: Influencing Our Planet and Re-engineering Our Actions, Taylor & Francis Group/CRC Press, Boca Raton, FL.

Badiru, Adedeji B., Nils Wagenknecht, Andreas Mertens, Olufemi Omitaomu (2024b), "Quantitative Systems Modeling for Critical Infrastructure Predictions in Climate Change: A National Defense Framework," *ASCE-ASME Journal of Risk and Uncertainty in Engineering Systems Part B: Mechanical Engineering*, March 2024, Vol. 10, Online at asmedigitalcollection.asme.org/risk. DOI: 10.1115/1.4063793

Doran, Eric, Sharon Bommer, and Adedeji Badiru (2022), "Integration of Human Factors, Cognitive Ergonomics, and Artificial Intelligence in the Human-Machine Interface for Additive Manufacturing," *International Journal of Quality Engineering and Technology, International Journal of Mechatronics and Manufacturing Systems*, Vol. 15, No. 4, 2022, pp. 310–330.

Maxwell, John C. (2024), *High Road Leadership: Bringing People Together in a World that Divides*, Maxwell Leadership/Simon & Schuster, New York, NY.

Peacock, Brian and Adedeji Badiru (2023), *Industrial Engineering in Systems Design: Guidelines, Practical Examples, Tools, and Techniques*, Taylor & Francis Group/CRC Press, Boca Raton, FL.

2 Industrial Engineering Foundation for Technology Efficiency and Effectiveness

TECHNOLOGY AND MAN

Our ultimate goal is to use technology efficiently and effectively. Industrial engineering (IE), by virtue of its diverse applicability, offers a viable pathway for achieving that goal.

Everything we do is a system. Thus, it is imperative that we commit substantive efforts to the design of systems. IE offers the best avenue for achieving sustainable efficiency, effectiveness, and productivity in systems in the work environment (Peacock, 2019, 2020, 2021). This is even more imperative when systems design is involved because of the multiplicity of factors involved in the interplay of people, machines, and tools. This is why the subdisciplines of human factors and ergonomics always feature prominently in any presentation of IE. That expectation is not different in this book.

As industrial engineers, we specialize in fitting the people, the process, and the tools together, using systems principles and techniques. Figure 2.1 aptly conveys this versatility of the discipline of IE in finding a better way to accomplish operational goals and objectives.

IE cares not only about making widgets but also about what widgets are made of, how they are made, where they are made, when they are made, and why they are made. IE optimizes the pathways toward products, services, and results. The framework used throughout the rest of this book capitalizes on the versatile framework presented in Figure 2.1. Badiru (2014, 2019) and all the references therein present the tools, techniques, and methods of IE that are applicable to this context. As conveyed in Figure 2.1, human factors and ergonomics are foundational to the ways of IE, as a people-oriented discipline.

On a daily basis, business, industry, academia, and government face challenges in human work and performance. Not only must work be done, it must be done productively, safely, profitably, and consistently, which is what IE facilitates, from a systems perspective. Thus, the premise of this book is to use engineering systems methodology for enhancing human work and performance. The engineering foundation for the book is based on the various tools, models, and techniques previously developed and implemented by Dr. Brian Peacock. In this book, the peacock tools and techniques are incorporated into the trademarked Design, Evaluation, Justification, Integration

DOI: 10.1201/9781032659220-2

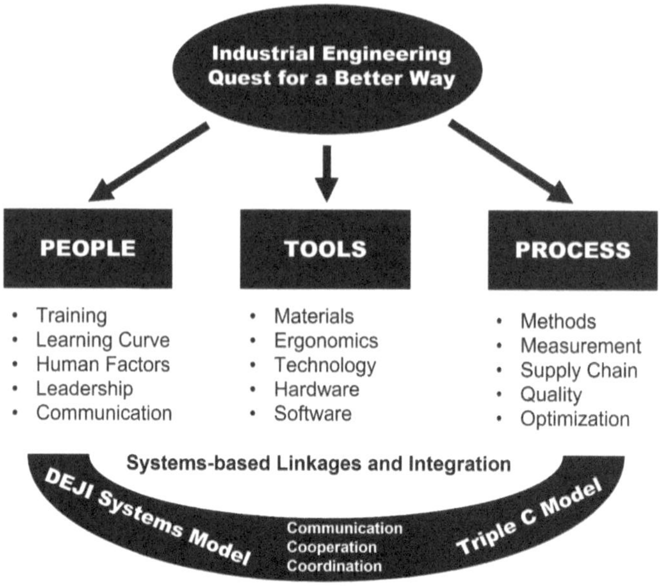

FIGURE 2.1 Industrial Engineering and the Quest for a Better Way.

(DEJI) Systems Model, providing a structured engineering process for design, evaluation, justification, and integration.

To understand the traction provided by IE, we have to recap the respective definitions of a system and the profession of IE.

A system is defined as a *collection of interrelated elements, whose collective output (in unison) is higher than the mere sum of individual outputs of the elements.*

> Industrial Engineering – A profession that is concerned with the design, installation, and improvement of integrated systems of people, materials, information, equipment, and energy by drawing upon specialized knowledge and skills in the mathematical, physical, and social sciences, together with the principles and methods of engineering analysis and design to specify, predict, and evaluate the results to be obtained from such systems.

Using systems-thinking approaches, IE makes direct contributions to teamwork, structural communication, critical thinking, smart design, impact assessment, operational literacy, and social responsibility. The above definition embodies the various aspects of what an industrial engineer does. As can be seen, the profession is very versatile, flexible, and diverse. It can also be seen from the definition that a systems orientation permeates the work of industrial engineers. Some of the major functions of industrial engineers involve the following:

- Design integrated systems of people, technology, processes, and methods
- Develop performance modeling, measurement, and evaluation for systems
- Develop and maintain quality standards for industry and business

- Apply production principles to pursue improvements in service organizations
- Incorporate technology effectively into work processes
- Develop cost mitigation, avoidance, or containment strategies
- Improve overall productivity of integrated systems of people, materials, and processes
- Recognize and incorporate factors affecting the performance of a composite system
- Plan, organize, schedule, and control production and service projects
- Organize teams to improve the efficiency and effectiveness of an organization
- Install technology to facilitate work flow
- Enhance information flow to facilitate smooth operations of systems
- Coordinate materials and equipment for effective systems performance

IE makes systems function better together with less waste, better quality, and fewer resources. The goal of every organization is to eliminate waste. Thus, the above definition is aptly relevant for everyone. (IE) can be described as the practical application of the combination of engineering fields together with the principles of scientific management. It is the engineering of work processes and the application of engineering methods, practices, and knowledge to production and service enterprises. IE places a strong emphasis on an understanding of workers and their needs in order to increase and improve production and service activities. IE activities and techniques include the following:

1. Designing jobs (determining the most economic way to perform work)
2. Setting performance standards and benchmarks for quality, quantity, and cost
3. Designing and installing facilities

The concluding chapter of this book highlights the DEJI systems model as a tool for IE, specifically for technology and systems integration, from a quality management standpoint. The model presents a structural template for systems design, evaluation, justification, and integration. Systems quality is used as the basis for illustrating how the model is applied across its four stages. Specific emphasis is placed on applying the model to technology management in the context of pursuing and tackling UN's SDGs (Sustainable Development Goals) and NAE's grand challenges.

INDUSTRIAL REVOLUTION TO TECHNOLOGY REVOLUTION

IE has a proud heritage with a link that can be traced back to the ***industrial revolution***. Although the practice of IE has been in existence for centuries, the work of Frederick Taylor in the early twentieth century was the first formal emergence of the profession. It has been referred to with different names and connotations. Scientific management was one of the original names used to describe what industrial engineers do. ***Industry***, the root of the profession's name, clearly explains what the profession is about. The dictionary defines industry generally as the ability to produce and deliver goods and services. The "industry" in IE can be viewed as the application of skills and cleverness to achieve work objectives. This relates to how human effort is harnessed innovatively to carry out work. Thus, any activity can be defined as "industry" because it generates a product; be it service or physical product. A systems view of IE encompasses all the details and aspects necessary for applying skills and cleverness to produce work efficiently. However, the academic curriculum of IE must change, evolve, and adapt to the changing systems environment of the profession.

It is widely recognized that the occupational discipline that has contributed the most to the development of modern society is *engineering* through its various segments of focus. Engineers design and build infrastructures that sustain the society. These include roads, residential and commercial buildings, bridges, canals, tunnels, communication systems, healthcare facilities, schools, habitats, transportation systems, and factories. Across all of these, the IE process of systems integration facilitates the success of the infrastructures. In this sense, the scope of industrial and systems engineering steps through the levels of activity, task, job, project, program, process, system, enterprise, and society. This handbook of industrial and systems engineering presents essential tools for the levels embodied by this hierarchy of functions. From the age of horse-drawn carriages and steam engines to the present age of intelligent automobiles and aircraft, the impacts of industrial and systems engineering (ISE) cannot be mistaken, even though the contributions may not be recognized in the context of the ISE disciplinary identification.

It is essential to recognize the alliance between "industry" and IE as the core basis for the profession. The profession has gone off on too many different tangents over the years. Hence, it has witnessed the emergence of industrial engineering professionals who claim sole allegiance to some narrow line of practice, focus, or specialization rather than the core profession itself. Industry is the original basis of IE, and it should be preserved as the core focus, which should be supported by the different areas of specialization. While it is essential that we extend the tentacles of IE to other domains, it should be realized that over-divergence of practice will not sustain the profession. The continuing fragmentation of IE is a major reason to compile a handbook such as this. A fragmented profession cannot survive for long. The incorporation of systems can help to bind everything together.

THE TYPEWRITER AND THE ADVANCEMENT OF COMMUNICATION

Notable industrial developments that fall under the purview of the practice of IE range from the invention of the typewriter to the invention of the automobile. Ergonomics and human factors provide the functional adeptness of the typewriter, from which we can discern the basis for the design of many modern products. Writing is a basic means of communicating and preserving records. It is one of the most basic accomplishments of the society. The course of history might have taken a different path if early writing instruments had not been invented at the time that they were. Below is the chronological technical evolution of the typewriter:

1714: Henry Mill obtained a British patent for a writing machine.
1833: Xavier Progin created a machine that uses separate levers for each letter.

1843: American inventor, Charles Grover Thurber, developed a machine that moves paper horizontally to produce spacing between lines.
1873: E. Remington & Sons of Ilion, NY, manufacturers of rifles and sewing machines, developed a typewriter patented by Carlos Glidden, Samuel W. Soule, and Christopher Latham Sholes, who designed the modern keyboard. This class of typewriters wrote in only uppercase letters but contained most of the characters on the modern machines.
1912: Portable typewriters were first introduced.
1925: Electric typewriters became popular. This made typeface to be more uniform. International Business Machines Corporation (IBM) was a major distributor of this product.

In each case of product development, engineers demonstrate the ability to design, develop, manufacture, implement, and improve integrated systems that include people, materials, information, equipment, energy, and other resources. Thus, product development must include an in-depth understanding of appropriate analytical, computational, experimental, implementation, and management processes.

OPERATIONAL EFFICIENCY

Lee and Johnson (2014) present a comprehensive treatment of operational efficiency, which is the basis for every systems design. The fields of engineering and management associate *efficiency* with how well a relevant action is performed, i.e. "doing things right", and *effectiveness* with selecting the best action, i.e. "doing the right thing." Thus, a firm is *effective* if it identifies appropriate strategic goals and *efficient* if it achieves them with minimal resources. This chapter focuses on *operational efficiency* or the ability to deliver products and services cost effectively without sacrificing quality. In this chapter, we investigate a firm's operational efficiency with both queueing models and productivity and efficiency analysis methods that identify maximum productivity and measure efficiency as a ratio of observed productivity to maximum productivity. The maximum productivity level serves as a benchmark for desired performance. The methods for analysis will vary depending on the level of analysis. For example, at the micro-level, we measure operational efficiency at points (machine, workstation, laborer) on the shop floor, whereas the macro-level might be at the firm, industry, or nation level. We begin by evaluating performance at the operational level and then apply productivity and efficiency analysis to aggregate performance at higher levels. The analysis of productivity and efficiency is associated with production economics which focuses on assessment and uses an aggregate description of technology to answer questions such as (Hackman, 2008):

- How efficient is the firm in utilizing its inputs to produce its outputs?
- Is the firm using the right mix of inputs or producing the right mix of outputs given prevailing prices?
- How will the firm respond to a price hike in a critical input?
- How efficient is the firm in scaling its operations?
- Has the firm improved its productive capability over time?
- How does the firm compare to its competitors?

The *strategic level* includes long-term planning issues such as make-or-buy decisions. The *tactical level* describes midterm actions that are done perhaps on a weekly or monthly basis, while the *operational level* emphasizes daily scheduling and shop floor control. Product Efficiency Acceptance (PEA) supports tactical-level decisions and is part of midterm production planning. PEA provides performance benchmarking and production guidance. It can also provide ex post analysis to quantify efficiency for complex production processes that use multiple inputs to generate multiple outputs, or ex ante analysis to suggest guidelines for resource allocation.

Absolute Operational Efficiency

Ideal benchmarks to measure efficiency are usually developed in a design laboratory under perfect operating conditions. However, it is not easy to identify the sources of efficiency loss between ideal performance and the best observed performance. For instance, in a manufacturing process operating in perfect conditions, one machine's ideal throughput is 100 units per hour, yet the actual throughput is 80 units per hour due to operator's skill, scheduling, etc. We can estimate an absolute operational efficiency (AOE) as:

$$\text{AOE} = \frac{\text{actual throughput}}{\text{ideal throughput}} = \frac{80}{100} = 0.8$$

Note that ideal benchmarks can be observed at the machine or process level but are almost never observed at the firm level. Thus, alternative metrics are beneficial in cases when ideal benchmarks are not observable.

RELATIVE OPERATIONAL EFFICIENCY

Relative operational efficiency (ROE) is the ratio of actual throughput compared to best observed throughput. Relative benchmarks are often used to measure efficiency because similar comparable machine, process, and firm, etc. are often easily identifiable. We estimate ROE by identifying the best observed performance in a data set of multiple operations performing the same task, for instance, a data set of multiple machines performing the same manufacturing process. We find that the best observed throughput is 90 units per hour, but machine A produces 80 units per hour. We can estimate the ROE of machine A as:

$$\text{ROE} = \frac{\text{actual throughput}}{\text{best observed throughput}} = \frac{80}{90} = 0.88$$

Best observed throughput is often determined by using historical performance data under the assumption if all conditions are unchanged, actual throughput should be equal to/or close to the historically best performance.

In the real world, a firm's resources are always limited. When a firm would like to provide a product or service, it must consume input resources to generate the output level. In this setting, operational efficiency is determined by the outputs produced as well as the input resources or costs consumed. Thus, we can define productivity and efficiency as:

$$\text{productivity} = \frac{\text{output}}{\text{input}}$$

$$\text{efficiency} = \frac{\text{productivity}}{\text{productivity of best pratice}}$$

In other words, productivity is the ratio of output level to the input level, and efficiency is the ratio of the current productivity level to the best practice productivity level. Best practice is defined as the largest productivity achievable.

The relationship between the output levels produced as input levels change is the production function.

PERFORMANCE INDEX FOR PRODUCTION

At the shop-floor level, queueing models provide a method for evaluating machine performance. In the model below, we use the notation M/M/1 to describe the inter-arrive process, the service process for a single-server queueing system. The first M indicates customer arrivals follow a Poisson (Markovian) Process, and the inter-arrival time is exponential distribution. The second M indicates the service time follows an exponential distribution. The 1 indicates there is a single server. We use two parameters to describe the M/M/1 queueing system. Let λ be the arrival rate and μ be the service rate. For example, if $\lambda = 2.5$ customers per hour, it means on average 2.5 individuals arrive every hour. Thus, $1/\lambda$ is the mean inter-arrival time and $1/\mu$ is the mean service time. The condition $\lambda < \mu$ is necessary for the system to be stable, i.e. for the queue to be finite in length. $\rho = \lambda / \mu$ is the probability the server is busy and p_0 is the probability the server is idle. p_i is the probability of the server with i customers. We use the following set of algebraic equations to analyze the queue's performance.

In the beginning, we want to know the stable probability p_0. To characterize a queueing system with the state transition between 0 and 1, a rate-balance equation between the arrival rate and service rate can be shown as:

$$\lambda p_0 = \mu p_1 \rightarrow p_1 = \left(\frac{\lambda}{\mu}\right) p_0 = \rho p_0.$$

Intuitively, an empty system needs 1 arrival to become state 1; a system with 1 customer needs 1 departure to become state 0. This idea is the foundation of the rate-balance equation.

Similarly, we can derive the rate-balance equation for state 1 associated with state 0 and state 2.

$$(\lambda + \mu) p_1 = \lambda p_0 + \mu p_2 \rightarrow p_2 = (1 + \rho) p_1 - \rho p_0 = \rho^2 p_0$$

We can also derive a general formula, $p_n = \rho^n p_0$, for the probability that there are n customers in the system (p_n).

We obtain p_0 since the sum of all probability p_n for $n = 1,\ldots,\infty$ must be equal to 1:

$$\sum_{n=0}^{\infty} p_n = p_0 \sum_{n=1}^{\infty} \rho^n = \frac{p_0}{1-\rho} = 1 \rightarrow p_0 = 1-\rho.$$

Thus, we derive the steady-state probability:

$$P[\text{server idle}] = p_0 = 1-\rho$$

$$P[\text{server busy}] = 1-p_0 = \rho = \lambda/\mu \text{ (also called the "utilization")}$$

$$P[\text{n customers in the system}] = p_n = \rho^n(1-\rho)$$

To derive the probability of n or more customers in the system:

$$\sum_{m=n}^{\infty} p_m = (1-\rho)\sum_{m=n}^{\infty} \rho^m = (1-\rho)\sum_{k=0}^{\infty} \rho^{n+k} = \rho^n(1-\rho)\sum_{k=0}^{\infty} \rho^k = \rho^n(1-\rho)\frac{1}{1-\rho} = \rho^n$$

$$P[\text{n or more customers in the system}] = \rho^n$$

$$P[\text{less than n customers in the system}] = 1-\rho^n$$

So far, the probability distribution of steady state is derived for a single-server queueing system. We can construct two indices to evaluate the queueing system's performance by asking:

What is the expected number of customers in the system/in the queue?
What is the expected time of a customer staying in the system/in the queue?

Let L be the expected number of customers in the system.

$$L = \sum_{n=0}^{\infty} np_n = \sum_{n=0}^{\infty} n\rho^n(1-\rho) = \sum_{n=0}^{\infty} n(\rho^n - \rho^{n+1}) = 1(\rho^1-\rho^2) + 2(\rho^2-\rho^3) + 3(\rho^3-\rho^4) + \ldots$$

$$= \rho + \rho^2 + \rho^3 + \rho^4 + \ldots = \rho(1+\rho+\rho^2+\rho^3+\ldots) = \frac{\rho}{1-\rho} = \frac{\lambda}{\mu-\lambda}$$

The expected number of customers in the queue, L_q, can be derived similarly. Note that we assume the customer being served is not in the queue, so n customers in the system means the queue length is $n-1$.

$$L_q = \sum_{n=1}^{\infty}(n-1)p_n = \sum_{n=1}^{\infty}np_n - \sum_{n=1}^{\infty}p_n = L-(1-p_0) = \frac{\rho}{1-\rho} - (1-(1-\rho)) = \frac{\rho}{1-\rho} - \rho$$

$$L_q = L - \rho = L\rho$$

Let W be the expected time spent in the system by a customer. Intuitively, it is equal to the expected number of customers in the system divided by arrival rate λ. The equation is:

$$L = \lambda W$$

This equation, or Little's law, defines the relationship between L and W. Similarly, $L_q = \lambda W_q$, where W_q denotes the expected time spent in the queue by a customer:

$$W = \frac{L}{\lambda} = \frac{L\rho}{\lambda\rho} = \frac{L_q/\lambda}{\rho} = \frac{W_q}{\rho} \rightarrow W_q = \rho W$$

$$W = \frac{L}{\lambda} = \frac{L_q + \rho}{\lambda} = \frac{\mu L_q + \lambda}{\lambda\mu} = \frac{L_q}{\lambda} + \frac{1}{\mu} = W_q + \frac{1}{\mu}$$

The relationship between W and W_q results because the expected time spent in the system is equal to the expected time spent in the queue plus the mean service time.

Above, Little's law is defined for a general queueing system. In a manufacturing system, Little's law is interpreted as the relationship among work-in-process (WIP), throughput (TH), and cycle time (CT):

$$\text{WIP} = \text{TH} \times \text{CT}$$

WIP is the number of unfinished units in the production system, TH is the number of finished products manufactured per unit of time, and CT is the amount of time the units remain in the production system. Given a fixed WIP, an inverse relationship characterizes TH and CT, i.e. an increase in TH will decrease CT.

Little's law is useful because it applies to a wide variety of production systems. Given a fixed TH, WIP and CT will maintain an almost linear relationship until the capacity limit is approached, but if WIP continues to increase, CT will deteriorate rapidly. When utilization approaches 100%, the increase of arrival rate λ will deteriorate WIP or CT. Thus, $\lambda > \mu$ implies the workstation is no longer stable. The typical performance metrics for queueing systems are utilization and throughput. We calculate utilization as:

$$\text{Utilization} = \frac{\lambda}{\mu} = \frac{\text{actual throughput}}{\text{theoretical (ideal) throughput}}$$

Given CT and the level of WIP, we use Little's law to calculate the M/M/1 system's productivity by dividing TH by 1. More complicated network analyses are possible with multiple processors linked in a network, Gautam (2012). Queueing theory can be used to calculate throughput, and productivity can be estimated by dividing throughput by the number of processors. However, all processors may not be identical and throughput will clearly be impacted by the underlying network structure. Further, the human component of operating machines adds additional complications and uncertainty that are difficult to capture in queueing models. Thus, production functions are useful for estimating complex systems or firm-level performance.

PRODUCTIVITY AND THE PRODUCTION FUNCTION

A production function $f(\mathbb{x})$ is the maximum outputs that can be achieved using input vector $\mathbb{x} = (x_1,\ldots,x_N)$ (Hackman, 2008). Outputs are units a firm generates, and inputs are the factors of production, or the commodities used in production. In economics, there are at least five types of factors of production: capital, labor, land, energy, and raw materials. We can analyze the performance of a firm's production system by using either the long-run production function or the short-run production function. In the short run, the factors can be divided into fixed factors and variable factors. Fixed factors are the factors that cannot be changed in the short run such as building and land, and variable factors are the factors that can be changed in the short run such as temporary workers. In the long run, all of the production factors are variable.

Theoretically, four properties characterize a production function (Chambers, 1988; Coelli, et al., 2005):

Nonnegativity: The production output is a finite, nonnegative, real number.

Weak Essentiality: The production output cannot be generated without the use of at least one input.

Monotonicity: Additional units of input will not decrease output; also called nondecreasing in $\mathbb{x}$.

Concavity: Any linear combination of the vectors $\mathbb{x}^0$ and $\mathbb{x}^1$ will produce an output that is no less than the same linear combination of $f(\mathbb{x}^0)$ and $f(\mathbb{x}^1)$. That is, $f(\lambda\mathbb{x}^0 + (1-\lambda)\mathbb{x}^1) \geq \lambda f(\mathbb{x}^0) + (1-\lambda)f(\mathbb{x}^1)$. This property implies the "law of diminishing marginal returns."

These properties can be relaxed to model-specific production behaviors. For example, monotonicity is relaxed to model input congestion (Färe et al., 1985, 1994), and concavity is relaxed to characterize an S-shaped production function (Frisch, 1964 and Henderson and Quandt, 1980).

Because of the fixed factors in the short run, the production function is characterized by monotonically increasing levels and diminishing returns, that is, increasing one variable factor of production will increase output levels at a decreasing rate while holding all others constant. The fixed factors limit the growth of the output. This is also called the law of diminishing marginal returns (product). Three concepts of production characterize a short-run production function:

Total product (TP): the total amount of output generated from the production system, $TP = y = f(x)$. Average product (AP): the average amount of output per unit input, $AP = f(x)/X$.

Marginal product (MP): the marginal change while adding one more unit of input, $MP = df(x)/dx$.

As a firm increases its input levels, the output levels also increase. The firm reaches point A, an inflection point, i.e. where the maximal MP is achieved. As inputs continue to increase, the single-input and single-output production function shows diminishing marginal product as it reaches the most productive scale size (MPSS). MPSS is the point on the production function that maximizes the AP (or productivity). Finally, input and output levels continue to increase until point B, beyond which input congestion occurs due to the fixed factors and negative MP.

All of the factors of production are variable in the long run. Consider production using multiple inputs. It is common practice to plot the relationship between two of the variables while holding all others fixed. Graphical analysis can show the relationship between the inputs x_n and x_m while holding the output fixed at the value y^0 and holding all other inputs fixed. The resulting curve is the input isoquant that gives all combinations of x_n and x_m capable of producing the same output level y^0. It is convex toward the origin if it satisfies all properties of the production function. For different output levels $y^2 > y^1 > y^0$, these isoquants form non-intersecting functions. The slopes of the isoquants are the marginal rate of technical substitution (MRTS) which measures the rate of using x_n to substitute x_m while holding the output level constant:

$$MRTS_{nm} = -\frac{\partial x_m(x_1,\ldots,x_{m-1},x_{m+1},\ldots,x_N)}{\partial x_n} = \frac{MP_n}{MP_m}$$

$$MP_n\,\partial x_n + MP_m \partial x_m = 0$$

where $x_m(x_1,\ldots,x_{m-1},x_{m+1},\ldots,x_N)$ is an implicit function indicating how much x_m is needed to produce the same output level given fixed levels of $x_1,\ldots,x_{m-1},x_{m+1},\ldots,x_N$. Thus, the rate of substitution of input m for input n along the isoquant is equal to the ratio of the marginal productivity of n relative to the marginal productivity of m. To remove the unit of measurement, the direct elasticity of substitution (DES) is the percentage change in the input ratio relative to the percentage change in the MRTS and quantifies the curvature of the isoquant:

$$DES_{nm} = \frac{d(x_m/x_n)}{d(MP_n/MP_m)} \times \frac{MP_n/MP_m}{x_m/x_n}$$

Next, we describe three typical production functions for a two-input case. Leontief production functions or fixed proportions functions describe production that occurs in fixed proportions, for example, cars that require wheels (x_n) and bodies (x_m). The mathematical form is $y = \min\{\beta_n x_n, \beta_m x_m\}$ and $\beta_n, \beta_m > 0$; the horizontal part of the isoquant indicates that an increase in x_n does not contribute to the output (y), and $MP_n = 0$ and $MRTS_{nm} = 0$, and that the vertical part of the isoquant indicates that an increase in x_m does not contribute to the output (y), and $MP_m = 0$ and $MRTS_{nm} = \infty$.

$MRTS_{nm}$ is not defined at the corner. Therefore, a Leontief production function is used to model production where there is no substitution between x_n and x_m, i.e. $DES_{nm} = 0$.

A linear production function assumes that inputs are substituted at a constant rate regardless of the level of either input or output. The mathematical form is $y = \beta_n x_n + \beta_m x_m$ and $\beta_n, \beta_m > 0$. The production function implies a constant rate of substitution, $MRTS_{nm} = MP_n / MP_m = \beta_n / \beta_m$, and also imposes perfect substitution between x_n and x_m, i.e. $DES_{nm} = \infty$.

COBB-DOUGLAS PRODUCTION FUNCTION

A Cobb-Douglas production function assumes that inputs are substitutable. However, consistent with the law of diminishing marginal productivity, additional inputs are needed to maintain the same output level as the mix of inputs becomes more skewed. The mathematical form is $y = \alpha x_n^{\beta_n} x_m^{\beta_m}$ and $\alpha, \beta_n, \beta_m > 0$; computational analysis shows that the production function is a smooth curve and convex toward the origin, and that $MRTS_{nm} = MP_n/MP_m = \alpha\beta_n x_n^{\beta_n - 1} x_m^{\beta_m} / \alpha\beta_m x_n^{\beta_n} x_m^{\beta_m - 1} = \beta_n x_m / \beta_m x_n$ decreases with respect to x_n. Thus, substitution exists in this production function and $0 < DES_{nm} < \infty$.

Production functions are convex toward the origin because the absolute value of the slope of the isoquant decreases while increasing x_n, thus $MRTS_{nm}$ also decreases. This is called the law of diminishing MRTS. The mathematical representation is $\frac{\partial}{\partial x_n} MRTS_{nm} < 0$.

In addition, if a proportionate increase in all inputs results in a less than proportionate increase in output, we say that the production function exhibits decreasing returns to scale (DRS). Alternatively, if increasing all inputs results in the same proportional increase in output, we say that it exhibits constant returns to scale (CRS). Finally, if the increase of all inputs results in a more than proportionate increase in output, we say that the production function exhibits increasing returns to scale (IRS). There are many reasons why firms may exhibit different returns to scale. For example, a firm may exhibit IRS if hiring more personnel allows specialization of labor, but the firm may eventually exhibit DRS if the firm becomes so large that management is no longer able to control operations. Firms that can replicate all aspects of their operations exhibit CRS. Operating at DRS would indicate decentralization or downsizing might be appropriate, whereas operating at IRS would indicate mergers, acquisitions, or other changes in organizational structure might be appropriate.

BENCHMARKING EFFICIENCY

IE analysts design the production function to define a benchmark to measure how efficiently production processes use inputs to generate outputs. Given the same level of input resources, inefficiency is indicated by lower levels of output. In a competitive market, if a firm is far from the production function and operates inefficiently, it needs to increase its productivity to avoid going out of business.

Production theory provides a useful framework to estimate the production function and efficiency levels of a firm in three ways: 1) using parametric functional forms in regression-based methods, e.g., Stochastic Frontier Analysis (SFA) (Aigner et al., 1977; Meeusen and van den Broeck, 1977), 2) using nonparametric linear programming methods, e.g. Data Envelopment Analysis (DEA) (Charnes et al., 1978; Banker et al., 1984) or 3) integrating regression and programming methods (Kuosmanen and Johnson, 2010).

STOCHASTIC FRONTIER ANALYSIS

Aigner and Chu (1968) use the logarithmic form of the Cobb-Douglas production function to estimate a deterministic frontier:

$$\ln y_k = x_k^{'}\beta - u_k$$

where $k = 1,\ldots,K$ and y_k indicates the single output of the firm k; x_k is a $I \times 1$ vector with the elements of logarithm inputs; β is a vector of unknown parameters; and u_k is a nonnegative random variable associated with technical inefficiency. Several methods can be used to estimate the parameter β, such as maximum likelihood estimation (MLE) or ordinary least squares (OLS) (Richmond, 1974). However, the Aigner and Chu method neglects statistical noise and assumes that all deviations from the frontier are a result of technical inefficiency. Therefore, Aigner et al. (1977) and Meeusen and van den Broeck (1977) proposed the stochastic frontier production function and introduced the random variable representing statistical noise as:

$$\ln y_k = x_k^{'}\beta + v_k - u_k$$

where v_k models the statistical noise using a symmetric random error. The function is bounded from above due to the stochastic variable $\exp\left(x_k^{'}\beta + v_k\right)$. To illustrate, we use a Cobb-Douglas stochastic frontier model with a single input variable:

$$\ln y_k = \beta_0 + \beta_1 \ln x_k + v_k - u_k$$

$$y_k = \exp\left(\beta_0 + \beta_1 \ln x_k\right) \times \exp\left(v_k\right) \times \exp\left(-u_k\right)$$

In this functional form, $\exp\left(\beta_0 + \beta_1 \ln x_k\right)$ is the deterministic component, $\exp\left(v_k\right)$ is the statistical noise, and $\exp\left(-u_k\right)$ is the inefficiency component. Lee and Johnson (2014) illustrate the deterministic frontier $y_k = \exp\left(\beta_0 + \beta_1 \ln x_k\right)$, the noise effect and the inefficiency effect of firm A and firm B. Firm A has a negative random noise component, whereas firm B has a positive noise random noise component. The observed output level $y_k = \exp\left(\beta_0 + \beta_1 \ln x_k + v_k - u_k\right)$ and the frontier output level (i.e. without the inefficiency effect) is $y_k^* = \exp\left(\beta_0 + \beta_1 \ln x_k + v_k\right)$. The observed output of firm B lies below the deterministic part of the frontier because the sum of the noise and inefficiency is negative. Since SFA estimates the inefficiency effects, we can define the output-oriented measure of technical efficiency (TE) by using the observed output over the frontier output:

$$TE_k = \frac{y_k}{\exp(x_k'\beta + v_k)} = \frac{\exp(x_k'\beta + v_k - u_k)}{\exp(x_k'\beta + v_k)} = \exp(-u_k)$$

This TE_k estimate shows the measure of the observed output of firm k relative to the frontier output of an efficient firm given the same input vector. This benchmarking with best practice provides the estimation of technical inefficiency.

We need to estimate the parameter vector β before calculating TE. Note that the model is complicated by the two random terms, v_i and u_i, where v_i is usually a symmetric error and u_i is a nonnegative term. The parameter β is estimated under the following assumptions:

$E(v_k u_l) = 0, \forall k, l$: uncorrelated
$E(v_k) = 0$: zero mean
$E(v_k^2) = \sigma_v^2$: homoskedastic
$E(v_k v_l) = 0, \forall k \neq l$: uncorrelated
$E(u_k^2) = \text{constant}$: homoskedastic
$E(u_k u_l) = 0, \forall k \neq l$: uncorrelated

Further, v_k and u_k are uncorrelated with the explanatory variables x_k. Note that $E(u_k) \neq 0$ since $u_k \geq 0$.

To estimate β, Aigner et al. (1977) assume $v_k \sim N(0, \sigma_v^2)$ and $u_k \sim N^+(0, \sigma_u^2)$, where v_k follows the independently and identically distributed (iid) normal distribution with zero mean and variance σ_v^2, and u_k follows the iid half-normal distribution which is a truncated normal distribution with zero mean and variance σ_u^2. This is called the "half-normal model" in SFA. Under these assumptions, the OLS estimator will provide consistent estimators of slope in β but a downward-biased intercept coefficient since $E(u_k) \neq 0$. Therefore, we use the MLE on the log-likelihood function with $\sigma^2 = \sigma_v^2 + \sigma_u^2$ and $\xi^2 = \sigma_u^2 / \sigma_v^2$:

$$\ln L(y \mid \beta, \sigma, \lambda) = -\frac{K}{2} \ln\left(\frac{\pi\sigma^2}{2}\right) + \sum_{k=1}^{K} \ln\Phi\left(-\frac{\varepsilon_k \xi}{\sigma}\right) - \frac{1}{2\sigma^2} \sum_{k=1}^{K} \varepsilon_k^2$$

where y is a vector of log-outputs, $v_k - u_k = \ln y_k - x_k'\beta$ defines a composite error term ε_k I, and Φ is a cumulative distribution function of the standard normal random variable. Finally, we use the iterative optimization procedure to estimate the coefficient β (Judge et al., 1985).

DATA ENVELOPMENT ANALYSIS

DEA is an optimization-based approach that imposes the axiomatic assumptions of monotonicity and convexity and the minimum extrapolation principle (MEP) (Banker et al., 1984). MEP identifies the smallest set that satisfies the imposed production assumptions and envelops all the data. Thus, DEA estimates a piecewise linear production function based on the observed data points. The producer on the SDH can decrease the input level without reducing the output level or decrease the output level without changing the input level. We measure the slack in inputs or outputs along the dashed line segments distinguishing Farrell efficiency measure (Debreu, 1951; Farrell, 1957) and Koopmans efficiency measure (Koopmans, 1951). The Farrell measure defines TE as the maximum radial reduction in all inputs consistent with equivalent production of output. The Koopmans measure states that it is impossible for a firm to increase any output without simultaneously reducing another output (or increasing any input). Note that after all inputs have been radially reduced, additional slack may still exist in some but not all inputs. Thus, a Farrell efficient firm may not be Koopmans efficient.

Here, we present the widely used Farrell measure. First, we introduce the linear programming technique to estimate the production function and production possibility set. Let $x \in R_+^I$ denote the inputs and $y \in R_+^J$ denote the outputs of the production system. We define the production possibility set as $T \equiv \{(x, y) : x \text{ can produce } y\}$. X_{ik} is the ith input resource, Y_{jk} is the amount of the jth production output, and λ_k is the multiplier for the kth firm. The following model defines the feasible region of the production possibility set $\tilde{T}$. This is called the variable return to scale (VRS) DEA model (Banker et al., 1984) because decreasing MP is observed along the frontier:

$$\tilde{T} = \left\{ (x, y) : \sum_k \lambda_k Y_{jk} \geq Y_j, \forall j; \sum_k \lambda_k X_{ik} \leq X_i, \forall i; \sum_k \lambda_k = 1; \lambda_k \geq 0, \forall k \right\}$$

We use the DEA estimator to measure the efficiency. We describe the input-oriented TE (ITE) as measured using the distance function $D_x(x, y) = \inf\{\theta \mid (\theta x, y) \in \tilde{T}\}$.

Input-oriented DEA efficiency model

$$\min_{\theta}\{\theta \mid \sum_k \lambda_k Y_{jk} \geq Y_j, \forall j; \sum_k \lambda_k X_{ik} \leq \theta X_i, \forall i; \sum_k \lambda_k = 1; \lambda_k \geq 0, \forall k\}$$

Output-oriented DEA efficiency model

$$\max_{\omega}\{\omega \mid \sum_k \lambda_k Y_{jk} \geq Y_j \omega, \forall j; \sum_k \lambda_k X_{ik} \leq X_i, \forall i; \sum_k \lambda_k = 1; \lambda_k \geq 0, \forall k\}$$

We calculate $\theta = 1/\omega$ from the output-oriented DEA efficiency model i to get an output-oriented TE (OTE), θ. $\theta = 1$ implies an efficient firm and $\theta < 1$ implies an inefficient firm. As an illustrative example (see Lee and Johnson, 2014), three firms, A, B, and C are located in an input space constructed by holding the output level constant at $y = \bar{y}$. The solid line is the piecewise linear efficient frontier estimated by DEA. Firms B and C are located on the frontier, but firm A is on the interior of the estimated PPS, $\tilde{T}$. Using the Farrell measure to estimate the TE shows that the inputs of firm A can be reduced radially. Point D is the intersection of the line segments $\overline{OA}$ and $\overline{BC}$. In fact, point D is a convex combination of firms B and C. We estimate Firm A's TE as:

$$TE_A = \theta = D_x(x_A, y_A) = \frac{\overline{OD}}{\overline{OA}}$$

The benefits of both SFA (stochastic frontier analysis) and DEA can be achieved using the nonparametric regression approach, StoNED. The first stage of StoNED uses convex nonparametric least squares (CNLS) proposed by Hildreth (1954) and extended by Hanson and Pledger (1976) to estimate a function satisfying continuity, monotonicity, and global concavity – the standard regularity conditions for a production function. To include both random noise and technical inefficiency, Kuosmanen and Kortelainen (2011) combine the CNLS piecewise linear production function with the composite disturbance term concept from SFA. Let $x_k \in R_+^I$ be an input vector, $y_k \in R_+$ be an output, and f be an unknown frontier production function satisfying continuity, monotonicity, and concavity. The regression model is:

$$y_k = f(X_k) + \varepsilon_k \quad \forall k = 1,\ldots,K$$

$$\varepsilon_k = v_k - u_k \qquad \forall k = 1,\ldots,K$$

$$y_k = [f(x_k) - \mu] + [\varepsilon_k + \mu] = g(x_k) + \vartheta_k \qquad \forall k = 1,\ldots,k$$

where ε_k is a disturbance term with $E(\varepsilon_k) = 0 \forall k$, $Var(\varepsilon_k) = \sigma^2 < \infty \forall i$ and $cov(\varepsilon_k \varepsilon_j) = 0 \forall k \neq j$. We formulate the CNLS problem as the quadratic program:

$$\min_{\alpha\beta\varepsilon} \sum_k \varepsilon_k^2$$

$$\begin{aligned} \text{s.t} \quad & \varepsilon_k = y_k - (\alpha_k + x_k' \beta_k) && \forall k = 1,\cdots,K \\ & \alpha_k = x_k' \beta \leq \alpha_h + x_k' \beta_h && \forall h,k = 1,\cdots,K \\ & \beta_k \geq 0 && \forall k = 1,\cdots,K \end{aligned}$$

where α_k and β_k are the coefficients characterizing the hyperplanes of the frontier production function f. Note that α_k and β_k are specific to each firm k. The objective function minimizes the sum of squared disturbance terms. The equality constraint defines the disturbance term as the difference between an observed output and an estimated output. The inequality constraints comprise a system of Afriat inequalities (Afriat, 1972), imposing the underlying frontier production function to be continuous and concave. The last constraints enforce monotonicity. Unlike DEA, CNLS uses all of the data points to estimate a production function, making it more robust to outliers.

The CNLS estimator of the production function, $\hat{f}(x)$, is generally not unique, but the fitted output values at observed inputs, $\hat{f}(x_k)$, are unique (Kuosmanen, 2008). In fact, given the fitted output values, it is possible to derive the tightest lower bound of the frontier production function as the explicit lower bound representor function:

$$\hat{f}_{min}(x) = \min_{\alpha,\beta}\{\alpha + x_k'\beta \mid \alpha + x_k'\beta \geq \hat{y}_k \forall k = 1,\ldots,K\}$$

where $\hat{y}_k = \hat{f}(x_k)$ is the fitted output value. Since the tightest lower bound $\hat{f}_{min}$ is a piecewise linear function satisfying continuity, monotonicity, and concavity, we can use it as the unique CNLS estimator of the frontier production function f.

StoNED uses a similar approach to SFA for modeling inefficiency and noise terms. Consider the composite disturbance term:

$$\varepsilon_k = v_k - u_k \qquad \forall k = 1,\ldots,K$$

where the same properties for v_k and u_k are assumed.

The composite disturbance term violates the Gauss-Markov property that $E(\varepsilon_k) = E(-u_k) = -\mu < 0$; therefore, we modify the composite disturbance term as:

$$y_k = [f(x_k) - \mu] + [\varepsilon_k + \mu] = g(x_k) + \vartheta_k \qquad \forall k = 1,\ldots,k$$

where $\vartheta_k = \varepsilon_k + \mu$ is a modified composite disturbance with $E(\vartheta_k) = E(\varepsilon_k + \mu) = 0$ and $g(x_k) = f(x_k) - \mu$ is an average production function. Since g inherits the continuity, monotonicity, and concavity, the CNLS method can find the estimator of the average production function g. We formulate the composite disturbance CNLS problem as:

$$\begin{aligned}
&\min_{\alpha\beta\vartheta} \sum_k \vartheta_k^2 \\
\text{s.t}\quad &\vartheta_k = y_k - (\alpha_k + x_k'\beta_k) && \forall k = 1,\cdots,K \\
&\alpha_k = x_k'\beta \leq \alpha_h + x_k'\beta_h && \forall k,h = 1,\cdots,K \\
&\beta_k \geq 0 && \forall k = 1,\cdots,K
\end{aligned}$$

where α_k and β_k are the coefficients that characterize the hyperplanes of the average frontier production function g. Note that the composite disturbance CNLS problem only differs from the CNLS problem, and the sum of squared modified composite disturbances is minimized.

To illustrate the StoNED estimator, 100 observations of a single-input single-output Cobb-Douglas production function are generated, $y = x^{0.6} + v - u$. The observations, x, were randomly sampled from a uniform [1,10] distribution, v was drawn from a normal distribution with standard deviation of 0.5, and u was drawn from a half-normal distribution with standard deviation of 0.7.

The second stage of StoNED uses the modified composite residuals, $\hat{\vartheta}_k \forall k$, to separate the technical inefficiencies and random noises by applying the method of moments (Aigner et al., 1977; Kuosmanen and Kortelainen, 2011). Assuming that technical inefficiency has a half-normal distribution, $u_k \sim \left|N\left(0, \sigma_u^2\right)\right|$, and that random noise has a normal distribution, $v_k \sim N\left(0, \sigma_v^2\right)$, the estimated standard deviation of technical inefficiency and random noise is:

$$\hat{\sigma}_u = \sqrt[3]{\frac{\hat{M}_3}{\left(\frac{2}{\pi}\right)\left(1 - \frac{4}{\pi}\right)}}$$

$$\hat{\sigma}_v = \sqrt{\hat{M}_2 - \left(\frac{\pi - 2}{\pi}\right)\hat{\sigma}_u^2}$$

where $\hat{M}_2 = \frac{1}{n}\sum_k \left(\hat{\vartheta}_k - \hat{E}(\vartheta_k)\right)^2$ and $\hat{M}_3 = \frac{1}{n}\sum_k \left(\hat{\vartheta}_k - \hat{E}(\vartheta_k)\right)^3$ are the second and third sample central moments of the modified composite residuals. Moreover, $\hat{M}_3$ should be negative so that $\hat{\sigma}_u$ is positive. Intuitively, the composite residuals should have negative skewness, reflecting the presence of technical inefficiency. We calculate the expected technical inefficiency by:

$$\hat{\mu} = \hat{\sigma}_u \sqrt{2/\pi}.$$

Given $\left(\hat{\alpha}_k, \hat{\beta}_k\right)$ from the CNLS problem, we write the unique StoNED estimator of the frontier production function as:

$$\hat{f}_{min}(x) = \min_{\alpha,\beta}\{\alpha + x_k'\beta \mid \alpha + x_k'\beta \geq \hat{y}_k \forall k = 1, \ldots, K\} + \hat{\mu}$$

where $\hat{y}_k = \min_{h \in \{1,\ldots,n\}}\left\{\hat{\alpha}_h + x_k'\hat{\beta}_h\right\}$. We obtain the unique CNLS estimator of the average frontier production function, $\hat{g}_{min}$, by using the tightest lower bound representor function with the fitted output values, $\hat{y}_k$. Recall that $\hat{y}_k$ is calculated from the representor function and $\left(\hat{\alpha}_k, \hat{\beta}_k\right)$. Therefore, we obtain the frontier production function by additively shifting the unique CNLS estimator of the average frontier production function upwards by the expected value of technical inefficiency.

Given $\hat{\sigma}_u$ and $\hat{\sigma}_v$, the method introduced in Jondrow et al. (1982) can estimate firm-specific inefficiency. Specifically:

$$\hat{E}(u_k \mid \hat{\varepsilon}_k) = -\frac{\hat{\varepsilon}_k \hat{\sigma}_u^2}{\hat{\sigma}_u^2 + \hat{\sigma}_v^2} + \frac{\hat{\sigma}_u^2 \hat{\sigma}_v^2}{\hat{\sigma}_u^2 + \hat{\sigma}_v^2}\left[\frac{\phi\left(\hat{\varepsilon}_k / \hat{\sigma}_v^2\right)}{1 - \Phi\left(\hat{\varepsilon}_k / \hat{\sigma}_v^2\right)}\right]$$

where $\hat{\varepsilon}_k = \hat{\vartheta}_k - \hat{\mu}$, ϕ is the standard normal density function and Φ is the standard normal cumulative distribution.

EFFICIENCY IMPROVEMENT ANALYSIS

Overall equipment effectiveness (OEE) is a time-based metric to assess productivity and efficiency, particularly for the semiconductor manufacturing industry (Ames et al., 1995; SEMI, 2000, 2001; de Ron and Rooda, 2005). The traditional single index metrics of productivity, throughput, and utilization do not allow easy identification of the root cause for reduced productivity. The OEE definition describes six standard equipment states:

Nonscheduled state: Equipment is not scheduled to be used in production, such as unworked shifts, weekends, or holidays (including startup and shutdown).

Unscheduled down state: Equipment is not in a condition to perform its intended function due to unplanned downtime events, e.g. maintenance delay, repair, change of consumables or chemicals, and out-of-spec input.

Scheduled down state: Equipment is not available to perform its intended function due to planned downtime events, e.g. production test, preventive maintenance, and setup.

Engineering state: Equipment is in a condition to perform its intended function but is operated to conduct engineering experiments, e.g. process engineering, equipment engineering, and software engineering.

Standby state: Equipment is in a condition to perform its intended function but is not operated; the standby state includes no operator available (including breaks, lunches, and meetings), no items available (including no items due to lack of available support equipment), and no support tools.

Productive state: Equipment is performing its intended functions, e.g. regular production (including loading and unloading of units), work for third parties, rework, and engineering runs done in conjunction with production units.

We define OEE as:

$$\text{OEE} = \frac{\text{theoretical production time for effective units}}{\text{total time}}$$

We decompose OEE into the following subcomponents: availability efficiency (AE), operational efficiency (OE), rate efficiency (RE), and quality efficiency (QE) (de Ron and Rooda, 2005):

$$\text{OEE} = \text{AE} \cdot (\text{OE} \cdot \text{RE}) \cdot \text{QE} = \text{Availability} \cdot \text{Performance} \cdot \text{Quality}$$

where

$$\text{Availability} = \text{AE} = \frac{\text{equipment uptime}}{\text{total time}}$$

$$\text{Performance} = \text{OE} \cdot \text{RE}$$

$$\text{OE} = \frac{\text{production time}}{\text{equipment uptime}}$$

$$\text{RE} = \frac{\text{theoretical production timefor actual units}}{\text{production time}}$$

$$\text{Quality} = \text{QE} = \frac{\text{theoretical production time for effective units}}{\text{theoretical production time for actual units}}$$

The availability captures the difference between machine breakdown and processing. Performance characterizes the production time and throughput. The quality is described by the yield metric which is typically driven by scrap, rework, defects, and reject types. In other words, OEE is a metric to estimate the efficiency of theoretical production time for effective units. In particular, the theoretical production time means the production time without efficiency losses. In addition, two popular indices can be integrated into the OEE framework: mean time between failure (MTBF), or the average time a machine operates before it fails, and mean time to repair (MTTR), or the average time required to repair a failed component and return the machine to operation:

$$\text{AE} = \frac{\text{equipment uptime}}{\text{total time}} = \frac{\text{equipment uptime}}{\text{equipment uptime} \times \dfrac{(\text{MTTR} + \text{MTBF})}{\text{MTBF}}} = \frac{\text{MTBF}}{\text{MTTR} + \text{MTBF}}$$

OEE has two practical benefits. First, we can use its subcomponents to identify bottlenecks and improve productivity. In general, machines with high utilization are typically the bottlenecks. Because bottlenecks can shift depending on the product mix, it is important for engineers to identify and release bottlenecks quickly to maintain high throughput levels. Note that utilization is a necessary condition for bottleneck identification, but it does not mean that all high-utilization machines are bottlenecks. If the processing time of each product is the same and the variation in the production line is low, a machine may have high utilization without affecting throughput. Second, we can use OEE to separate the machine's status into regular operating conditions and down. The availability level quantifies the time used for production. A lower throughput is sometimes the result of low availability rather than poor performance. Thus, OEE decomposition helps with machine diagnosis and productivity improvement.

LEAN MANUFACTURING AND TECHNOLOGY

Lean manufacturing has its roots in the manufacturing processes developed by Henry Ford in the 1920s. The Ford Motor Company increased its revenue during the post-World War I depression by developing assembly line methods and eliminating activities that were either unnecessary or did not add value to the cars produced. Toyota coined the name and the concept of lean manufacturing in its production system in the 1980s and also developed additional supporting methods and concepts such as the just-in-time (JIT) system (Ohno 1988a, 1988b). We call a production system "lean" if it produces the required output levels with minimal buffering costs.

In fact, the only time a machine adds value is when it processes a part. Industrial analysts use a Gantt chart to visualize processing time, transportation time, and wait time. Note that loading products into tools is handling, not processing, and thus a non-value-adding activity. Most of the processing time of a product is waiting and non-value-adding activities. Smith (1998) proposed a manufacturing performance index called manufacturing cycle efficiency:

$$\text{Manufacturing Cycle Efficiency} = \frac{\text{Value} - \text{adding time}}{\text{Total cycle time}}$$

He pointed out that this index is often less than 1 percent in practice, meaning that firms usually waste resources performing non-value-adding activities. The basic philosophy of lean manufacturing is to eliminate waste by buy(ing) only enough material to fit the immediate needs of the production plan, considering the transportation of resources.

Below, we describe the three main principles of lean manufacturing:

Waste elimination
Continuous flow
Pull production system

Waste elimination reduces all forms of waste in the manufacturing process. Continuous flow smoothes and balances the production flow. Pull production system, or "make-to-order production" allows a firm to produce units only when it receives an order. There are four steps to implementing lean manufacturing:

Eliminate waste: 7 types of waste are identified and need to be eliminated
Use buffers: Build up, adjust, and swap buffers to manage for variability
Continuous improvement: A commitment to productivity improvement
Reduce variability: Identify and reduce internal and external causes

A firm can allocate resources dynamically and switch buffers to manage internal or external variability. Internal variability results from uncertain processing times, setups, machine breakdown, yield loss, rework, engineering change orders, etc., and external variability results from demand fluctuation, customer change orders, supplier uncertain delivery, etc. Lean manufacturing uses three buffers: inventory, capacity, and time. Inventory hedges against uncertain demand. Capacity is somewhat flexible due to hiring/layoffs of temporary workers, adjusting overtime, or outsourcing some activities. Time coordinates supply chain or manufacturing activities.

The benefits of lean manufacturing include:

Productivity improvement
Total manufacturing time saved
Less scrap
Lower inventory

Quality improvement
Plant space saved
Better labor utilization
Lower production cost, higher profits, and wages
Shorter CT: Make-to-order vs make-to-stock
Safety of operations

PRODUCTION WASTE ELIMINATION

Womack and Jones (2003) describe seven types of "muda," or waste, in a production system:

Transportation: Move products or materials that are not being processed between workstations, or between supplier and customer.
Inventory: Hold excess inventory of raw materials, WIP, or finished units.
Motion: Worker or tools move more than necessary, such as picking, sorting, stacking, or storing parts.
Waiting: Wait for upcoming tools, materials, parts, or for the next production step.
Overproduction: Generate excess products beyond the demand level.
Overprocessing: Working more than is necessary because of poor tool or product design.
Defects: Cost of poor quality such as rework, scrap, inspection, and repair.
In general, all seven types of waste described above belong to the category of non-value-adding activities.
Continuous flow, or the series of continuous and smooth processes, is the second principle. Each production step performs only the jobs necessary for the next step. Workstations do not hold unnecessary WIP and materials that block incoming and downstream flows.

SINGLE-MINUTE EXCHANGE OF DIE (SMED)

Single-minute exchange of die (SMED), or "Shingo", can significantly reduce setup time and improve productivity. Long setup time leads to a small number of setups, larger batch sizes, larger WIP inventories, and poor process flow. SMED divides the setup time into internal and external activities. An internal activity is one that can only be done when the machine is stopped such as multi-chambers adjustments; an external activity is anything that can be performed before or after the setup without stopping the machine, such as pre-heating of raw material. To achieve a quick setup and changeover of dies, SMED recommends reducing internal setup time or converting internal activities to external activities.

PRODUCTION LINE BALANCING

Line balancing is a typical problem of the assembly system design in IE (Nof et al., 1997). To compensate for demand fluctuations, the goal is to organize tasks into different groups, with each group taking the same amount of time. The line balancing problem is an NP-hard problem (Garey and Johnson, 1979); thus, heuristic methods are usually applied to provide good solutions. Helgeson and Birnie (1961) proposed a heuristic method called the ranked positional weight technique. This heuristic is a task-oriented technique considering the combination of precedence relationships and task processing time. Three steps are applied in this algorithm. Calculate the positional weight (PW) of each task using the processing time (PT) of the task plus the processing time of all tasks having this task as a predecessor.

RANK ORDERING OF TASKS

Assigns tasks to workstations sequentially in the ranked order, given the precedence relationships and CT constraint. If the CT is 10 minutes for each workstation, we calculate the minimal number of workstations according to the sum of the 8 task times over the CT, that is, 38/10 = 3.8 and round up to 4. However, this minimum number does not consider the precedence constraints. Thus, we use the ranked PW technique for line balancing. We find that the required number of workstations is five and the total idle time is 12, both of which tend to increase at downstream stations. The smoothing can be done by product type or by volume: both are quite efficient and can bring substantial efficiencies and savings. Note that a smoothed and continuous flow can be reviewed from a firm's internal production or its supply chain. The benefits include:

- Enhance flexibility by reducing batch size to accommodate changes in product mix or demand fluctuation.
- Reduce material, WIP, and inventory levels since there is no severe over- or under-production.
- No bottlenecks because of similar burdens for each workstation.
- Enhance loyalty and commitment to the firm, i.e. a stable workforce without temporary labor.
- Shorten changeover and setup times to reduce machine idleness.

PULL PRODUCTION SYSTEM

Push systems release work without consideration of system status and hence do not have an inherent limitation on WIP. The work is released based on a schedule of demand and controlled release rates, typically referred to as a due-date-driven production system. A pull system developed by Toyota releases work based on the status of the systems and has an inherent WIP limitation. The system authorizes work releases based on system status and controls WIP level. It is an order-driven production system (Hopp and Spearman, 2004). There are two techniques in the customer-pull production system: JIT and kanban. JIT attempts to reduce inventory, holding costs, and WIP using small lot size or even single unit processing. A "kanban" is a signboard or visual for realizing JIT and often leads to significant quality improvement. The advantages of using a pull production system include the following:

- Reduce WIP and CT: Limit releases into the production line.
- Improve quality: Short queues allow errors to be identified quickly and shut down the production line to correct the problems.

- Reduce cost: Switch the control from release rate to WIP level and reduce WIP progressively.
- Logistical benefits: Less congestion, easier control, and WIP cap control.
- Kanban provides for efficient lot tracking and predetermines WIP level by the number of kanban.

In fact, based on Little's law, $WIP = CT \times TH$, given the same rate of throughput, reducing the WIP level will lead to a reduction in CT. Thus, a pull production system reduces CT by controlling the WIP level. For further study of the pull system, see Ohno (1988a; 1988b), Liker (2004), and Nahmias (2009).

DIGITAL DIMENSION DISRUPTION

Technology application provides a means of getting things done to achieve specific goals of efficiency, effectiveness, and productivity. These are all within the operating alleys of IE. Although IE is often tied to production processes, the versatile tools and techniques of IE are applicable to a variety of operational scenarios. One good example is national security, which involved a systems framework of people, technology, and process, just as we might find in a production facility. Rybeck et al. (2017) present an account of a disruptive technology in national security, as recounted in the paragraphs and sections that follow.

The digital dimension is simultaneously enhancing and disrupting the fabric of life in every society where modern, *informatized* technology is present. The slow-motion collapse of parts of the 20th century's legacy is now accelerating in ways that likely will usher in a monumental realignment of societal institutions, methods of business, and fundamental ideas about national security. This realignment will, of necessity, change the frameworks within which America provides for its security, including how it acquires the goods and services it uses in that effort.

The U.S. National Security Enterprise (NSE or Enterprise) has not yet grasped, as evidenced in budget priorities, what it means to live in a world where the threats reside at considerable distance, at scales beyond our imaginings, and at speeds that cannot be easily comprehended. Information technology has penetrated all aspects of our lives and the *informatized threat* is a clear and present danger. The People's Republic of China has penetrated our defense supply chain, North Korea has exposed our corporate vulnerabilities, and Russia has threatened our social cohesion. From a national security perspective, informatized threats are by no means limited to the military or intelligence domains.

Yet, the Enterprise thus far has followed predictable, requirements-driven, program-oriented constructs that attempt to "normalize" responses, which subdivides the problem too early and misjudges its scale. What will it take to achieve enough common understanding to impel action? What will it take to align the NSE, its allies, and its partners to take effective, coordinated, and coherent countermeasures to maintain peace (when possible)?

The NSE needs an infusion of enterprise engineering originating within its most senior levels to establish new rules of engagement that match the emerging threat. *Informatized conflict* redefines the battlespace and demands a comprehensive and coherent response. Success depends on the active engagement of the entire diplomatic, economic, and military arsenal. This article adopts the best current, unclassified, holistic view of an informatized era vision for the Enterprise.

THE INFORMATIZED ERA

The word "computer" originally meant a person who did computation, like the clerks in 17th-, 18th-, and 19th-century brokerages. Computing machines (what is now meant when anyone says "computers") emerged in the mid-20th century as a quantum leap forward in how humans did calculations and searched for information. In the past quarter century, human use of computers has changed fundamentally, but common terminology has not kept pace with reality.

Society has become almost wholly dependent on informatized systems. As part of the creative destruction/evolution that drives capitalism, the pre-informatized infrastructure has been destroyed, but societal processes, especially those of government, defense, and the law—are still those of the pre-informatized world, a world that is rapidly going out of existence. The world is using digitized, sharable information, transitioning from one-way, single-supplier siloed, one-function stovepipes to interactive ecosystems where software is orchestrating the movement of goods and services, the making of decisions, and impacting the way humans live. In just one generation, every industry has come to depend on interactive real-time decision-making.

A careful look at what has changed in the transition from the computer era to the informatized era reveals a qualitatively new infrastructure that matured during the past twenty years. Distance and time are compressed to the point where an adversary's geography is not decisive (or, in many cases, even discernable), and the pace of action can be so fast that it defies normal human cognition. Most U.S. citizens can identify aspects of this new infrastructure such as broadband connectivity, massive availability of compute power on a global basis via the cloud, and the advent of big data. However, the implications of the changes brought by informatization have not broken through to the thinking guiding the highest levels of the U.S. Government.

The informatized era's new infrastructure is distinguishing itself by freeing increasingly mercurial data to move around the world – from place to place, from purpose to purpose – to feed previously unimagined analytics. Indeed, the nature of data is, itself, undergoing a fundamental change. The terms "bespoke data" (from the British term for custom-tailored) and "by-product data" highlight the difference between data created in the old pre-computer and computer worlds and data created by or in the new informatized world.

Bespoke data are made by a human using measurement tools, like much of traditional intelligence, created to answer a known question. By-product data are incidentally created by machine operations, like the geolocation data dropped by smartphones, and are then available for other use. By-product data are growing exponentially as a primary feature of the informatized era and are only in the infancy of technology exploitation by the NSE.

All informatized systems are essential to our national security irrespective of geography or commercial or government origins. Informatized conflict includes all national security-relevant activity, both kinetic and non-kinetic, whether it is commonly understood by practitioners as being in that context or not. For example, private commercial transactions are often conducted by their participants as if they had no national security implications. But all serious analysts recognize the indispensability of our critical infrastructure, including the electronic systems that facilitate commerce.

As anyone with a smartphone knows, the digital dimension is now integral to every aspect of business and societal interaction on a global scale. Viewed through the lens of informatized conflict, the "information technology" (IT) concept clearly fails to capture the full impact of the digital dimension on our world. The concept harkens back to the now-distant days when IT was a sequestered, relatively unimportant, compartment of our world. Chief Information Officers (CIOs) reported to Chief Financial Officers (CFOs) because Chief Executive Officers (CEOs) pigeonholed computers as simple aids to accounting.

While many summarize current threats under the term "cyber," a concept that points to everything digital, the terms "digital" and "cyber" are insufficient to capture the current threat dynamic. Cyber, for example, has usefully come to point specifically at computer network operations (CNO) but fails to capture the digital dimension as a whole. CNO's commonly described sub-divisions, computer network defense (CND), computer network exploitation (CNE), and computer network attack (CNA), encompass only a subset of the digital foundations on which modern life is being built.

ALIGNMENT AND INFORMATION FUSION

America entrusts its frontline national defense leadership to the Department of Defense (DOD) and the Intelligence Community (IC), two interconnected but separate chains of command. These entities are chartered to deliver kinetic and non-kinetic capabilities. Only the Commander-in-Chief (POTUS) controls both. As hard as it is for POTUS to exert Commander's intent, Congress faces even greater impediments when it attempts to prompt changes to how DOD prosecutes its mission. For example, Goldwater–Nichols (the U.S. Department of Defense Reorganization Act of 1986) demanded jointness in our military, and Clinger–Cohen (the Information Technology Management Reform Act of 1996) demanded IT rationalization. Neither the White House nor the Congress has directed DOD or the IC with sufficient clarity to guide execution of these for the 21st century.

DOD and IC systems are compartmentalized and often impervious to improvement with industry best practices. Every unit in the IC and DOD is sub-dividing the Enterprise problem and producing their own examples of this poorly aligned and tightly coupled approach. A massive array of programs and projects have been given carte blanche to operate using proprietary systems, creating processes that, while often narrowly effective, are impervious to improvement by informatized standards. In addition, the leaders of these programs are incentivized based on quick wins and continued resource growth, but these small pockets of capability do not add up to an Enterprise solution.

The vision centers around intelligence, surveillance, and reconnaissance (ISR) with all projections of national power. This vision has not been translated into a full-blown strategy and does not yet represent the NSE reality. It does, however, provide a strong basis for the fusion of command, control, communications, computers, intelligence, surveillance, and reconnaissance.

This vision of an informatized era "to be" depicts the alignment as it is required at the top and center of the Enterprise. Subsequent to that alignment, an unlimited number of loosely coupled implementations at the edge can then seamlessly connect and interoperate. This "tightly aligned/loosely coupled" engineering approach has been successfully applied at the Enterprise level in the private sector to guide foundational, internet-dependent initiatives. In less than three decades, for example, this approach has proven itself to be the most effective way for informatization to transform global enterprises, including Walmart, Netflix, and Google. It explains not only how the internet works but is ideally suited to support "innovation at the edge" for American warfighters.

Jointness in the informatized era needs to refer not only to the combined efforts of our armed services but also to the unified actions of DOD, the IC, and other stakeholders – and their ever-shifting alliances, whose efforts combine in pursuit of national security with all the instruments of national power. And fusion will need to combine data, data science, and data services to achieve the security objectives first outlined by the bipartisan 9/11 Commission.

Informatized fusion thus describes the new core competence that the NSE must develop to prevail in informatized conflict. The Chinese and the Russians have already adopted their variants of informatized fusion as guiding strategies. As a democracy, however, the United States requires popular understanding and support to pursue this strategy. Fortunately, the United States variant can maintain its comparative advantage by drawing on inherent American strengths – namely constitutionally protected rights as well as checks and balances built into three branches of government, private sector competition, the rule of law, and multi-ethnic diversity.

In the words of former Director of the Central Intelligence Agency Michael Hayden, America needs to balance "unity of effort," that is, tight alignment with "autonomy of action." That is, loose coupling. This new, agile, non-stove-piped approach to national security-related actions would allow asynchronous, near real-time intervention outside today's cumbersome processes. This vision is often cited in non-authoritative documents, but it has not yet been translated into a clear Commander's intent, Congressional intent, or the guiding National Strategy, nor has it been realized. Unfortunately, if America stays on its present course, it is not likely to get there. Now is the time to exploit a "tightly aligned and loosely coupled" strategy to fortify the NSE.

THE PURSUIT OF MISSION SUCCESS

Mission success is achieved only through authorizing initiatives at sufficient altitudes to match their charters and assigning responsibility to executives of sufficient gravitas. Informatization era challenges have their roots in the technology arena, but business-as-usual technological solutions alone will not address these challenges.

Decisionmakers and influencers from across the executive and legislative branches, with the support of the American public, will have to consider, adopt, and develop a joint 21st-century vision to realize the benefits of this digital technology reorientation. Champions are the only ones eligible to align and mobilize in the service of jointness, as redefined here to include the entire NSE.

Government governance and budget, mission execution, and technology elements perform functions analogous to their three familiar private sector equivalents, that is, the CEO team, the Chief Operating Officer (COO) team, and the CIO team. These three mission-critical teams combine to form the NSE and fulfill its mission. Any mission-critical team can initiate Enterprise-level innovation, but it is the joint action of all three together that delivers the Enterprise-level benefits.

The differences between the government's organization and the private sector are useful in understanding why commonsense solutions and efficiencies adopted almost universally in the private sector cannot be easily adopted by the government. Informatized fusion as a joint strategy would implement mechanisms for aligning all three mission-critical areas, expedite Enterprise-level solutions, and incorporate appropriate checks and balances into the decision-making process.

Ultimately, the Commander-in-Chief and Congress will need to mobilize the three mission-critical teams to meet the challenge of the digital dimension. To some observers, this will look like reprogramming, to others, it will present itself as major changes to mission processes, and for others, it will appear as technology transformation. To all those involved, however, it will reflect unprecedented alignment. This technology fusion demands cross-functional experience to fully accommodate their counterparts' frames of reference, demands, or "battle rhythms." Only a few, exceptional individuals in the government possess the required competencies, namely, vision of the end game, cross-functional credibility, and maturity based on experience with sustained and disciplined innovation at the highest levels, to galvanize support and align stakeholders around the mission. The champions of this strategy will require a Senior Executive Technical Review and be empowered to act on its findings. At the operational level, these champions will have to:

- Articulate a full-blown informatized fusion vision that matches the task and continually reminds everyone who will listen why the larger initiative is being undertaken.
- Align the vision's mission/business case (with quantitative and qualitative analysis of its risks and rewards) with its concept of operations and reference architecture.
- Arrange sufficient and sustained funding for all key elements of the initiative. Weak organizations use the mission/business case only to justify initial funding. Strong ones see a persistent, living mission/business case as a primary tool for guidance and for ensuring the delivery of promised benefits.
- Sequence activities based on announced priorities and predecessor/successor relationships to make sure benefits are delivered as promised. Only by delivering a no-kidding "without this…" list can a champion confront stakeholders with the stark reality of what it will take to achieve the benefits the champion presents in the corresponding and contingent "…you do not get that" list.
- Prioritize and communicate realistic expectations.

- Empower and incentivize executives at all levels when they enable shared, Enterprise-focused mission capabilities, and disincentivize silo-oriented approaches.

The right champions will know how to use a rigorous "mission/business case" to sustain alignment among the three mission-critical teams and to sustain bipartisan support. They will need to alter the rules of engagement under which the entire NSE conducts its business.

Fortunately, the mission benefits are so powerful and the cost savings so dramatic that a coherent and well-supported mission/business case at the informatized NSE level could overcome the entrenched interests who can be expected to fight it with all the tools at their disposal. Getting this right can unleash incredible growth and innovation. The potential may be compared with the 19th-century commitment to build railway lines with a consistent gauge (the distance between the rails), which was an essential step in the growth of an unimodal, continental economic engine.

Many of the new rules of engagement require changes in processes where the NSE is employing 18th-, 19th-, and 20th-century acquisition methodologies to solve contemporary, informatized problems that are mutating at an ever-increasing pace. Historic acquisition methodologies are not up to the current challenges, diminishing the NSE's ability to keep up with, let alone get ahead of the rapidly rising, digitally driven, innovation curve.

In the computer age, the NSE sought unique stand-alone things. While sometimes extraordinary, these solutions had pre-defined and relatively fixed capabilities, making them ill-suited to adapt to the changing needs of the stakeholders. In the informatized age, the focus has shifted toward integrated capabilities and solutions built on commodity technology. Even though these new informatization-aware systems are driven by specific missions, their capabilities are built to relentlessly adapt with the ever-changing needs of NSE-wide stakeholders.

Government champions, as described in this article, alone have the authority to prosecute informatized fusion and all it implies. Only they can move the biggest rocks, the Federal Acquisition Regulation (FAR) and Defense Federal Acquisition Regulation (DFAR), so that the NSE can coordinate – at the digital dimension level – procurement and deployment of virtually everything. With strong leadership, the NSE can rapidly transition from acquiring extraordinary things that confer relatively fixed capabilities to open-ended, increasingly extraordinary capabilities built using commodity things. The U.S. Air Force Space Command has already begun to shift from buying rockets (things with capabilities) to buying launches (capabilities).

In the absence of fully engaged champions, the NSE routinely avoids discussion of the cross-cutting capabilities on which informatized fusion depends. Lower-level government employees are not empowered or incentivized to operate at the scale or scope required to make the needed changes in either process or procurement. They are left waiting until aligned senior executive champions intervene to exercise their extraordinary and non-routine authority, changing the rules of how business is conducted. Until then, lower-level employees are reduced to reporting classic quick wins and low-hanging fruit. Until their boss' bosses make the tough choices and substantial investments needed for informatized fusion, the oft-touted mission benefits will remain elusive.

To ensure that the results of the champions' actions endure, this article looks to enterprise engineering, a discipline that makes practical application of systems engineering at the organization level, directing a venture in its entirety as a system-of-systems. It considers every aspect of the Enterprise, including business processes, information flows, material flows, organizational structure, and the human condition.

The USA Constitution represents one of the most successful and earliest examples of enterprise engineering. To ensure that the NSE has the resiliency to informatized change that gives it a lifespan comparable to that of the Constitution, the NSE needs an infusion of enterprise engineering originating in the most senior levels, establishing new rules of engagement that recognize the world is now irreversibly informatized.

The history of successful reengineering of processes within the national security arena has almost invariably been associated with mission process owners who were empowered to make the necessary changes. A good, though all-too-rarely-remembered, example was provided by Admiral Hyman Rickover, the father of the Navy's nuclear propulsion program. Because Rickover was so widely respected and because his authority was so significant, he was able to serve the NSE as an invaluable counterweight to the contractors who were building the ships, ultimately forcing the adoption of the standardized solutions required to achieve Enterprise-level alignment.

Rare exceptions only prove the rule: wherever process ownership is unassigned – as it is throughout most of the NSE on most national security processes – process improvement is left homeless, without adequate guidance and context. Advanced Research Project Agency (ARPANet) – the defense network that became the basis of the internet – demonstrated a means of exerting sufficient guidance and control to enhance the likelihood for success without stifling innovation or slowing the pace of change. ARPANet offered unprecedented connectivity and revolutionized information architecture. Here, the structure (packets in defined forms), flow (transmission), and management (orchestration) of information were transformed into what we all now recognize as the underlying foundation on which the modern internet is built.

Enterprise engineering has always required so much more than just managing the underlying technology. Whether dealing with the internet or the electrical grid, the private sector had to work with the public sector to set the standards. Subsequently, all enterprises (public and private) had to make major investments to adapt their business practices to take advantage of the new infrastructure (Gililland et al. (1995), Kuosmanen and Kortelainen (2012).

History shows that establishing foundational alignment cannot be accomplished through business-as-usual channels. Extraordinary interventions by the most senior executives – who, under business-as-usual conditions, typically have little involvement with infrastructure – was what proved decisive. Only after alignment was achieved through regularizing the structure, flow, and management of information could the work of adapting systems for exploiting that infrastructure be delegated. In the case of informatized fusion (combining cloud, mass analytics, and the projection of national power), the NSE will need to align around changes in the structure, flow, and management of information to begin what will be an ongoing process.

The NSE's current unaligned objectives, budgets, programs, policies, and procedures limit successful examples of enterprise engineering to isolated islands. Only a fallacy would suggest that such disparate efforts would produce informatized fusion. The nation, simply, cannot afford to wait until adversaries inflict catastrophic damage before it strategically aligns and takes the steps that it already knows are needed. In advance of the unthinkable, can a nation do what it takes to provide for the common defense in this era of information-technology-driven conflict?

OE can be measured and improved using the approaches described in this chapter. Today, many manufacturing firms define a metric for efficiency and concentrate on operational improvement activities to increase it. The specific approaches developed to identify best practice performance or to determine if a particular activity adds value are often product- or industry-specific. However, the evolution of new and global industries will require more sophisticated efficiency analysis techniques and metrics. IE tools and techniques could help with this sort of national-level operational challenge.

REFERENCES

Lee, Chia-Yen and Andrew L. Johnson (2014), "Operational Efficiency," Chapter Two in *Handbook of Industrial and Systems Engineering*, Adedeji Badiru, editor, pp. 17–44, Taylor & Francis Group/CRC Press, Boca Raton, FL.

Badiru, Adedeji B. (2019), *Systems Engineering Models: Theory, Methods, and Applications*, Taylor & Francis Group/CRC Press, Boca Raton, FL.

Badiru, Adedeji B., editor (2014); *Handbook of Industrial & Systems Engineering*, Second Edition, Taylor & Francis Group/CRC Press, Boca Raton, FL.

Afriat, S. N. (1972). Efficiency estimation of production functions. *International Economic Review* 13 (3): 568–598.

Aigner, D. J. and S. F. Chu (1968). On estimating the industry production function. *American Economic Review* 58: 826–839.

Aigner, D., C. A. K. Lovell, and P. Schmidt (1977). Formulation and estimation of stochastic frontier production function models. *Journal of Econometrics* 6: 21–37.

Banker, R. D., A. Charnes, and W. W. Cooper (1984). Some models for estimating technical and scale inefficiencies in data envelopment analysis. *Management Science* 30(9): 1078–1092.

Chambers, R. G. (1988). *Applied Production Analysis: A Dual Approach*. New York: Cambridge University Press.

Charnes, A., W. W. Cooper, and E. Rhodes (1978). Measuring the efficiency of decision making units, *European Journal of Operational Research* 2(6): 429–444.

Coelli, T. J., D. S. Prasada Rao, C. J. O'Donnell, and G. E. Battese (2005). *An Introduction to Efficiency and Productivity Analysis*. (2nd ed.). New York: Springer.

Debreu, G. (1951). The coefficient of resource utilization. *Econometrica* 19: 273–292.

Färe, R. S., S. Grosskopf, and C. A. K. Lovell (1985). Technical Efficiency of Philippine Agriculture. *Applied Economics* 17: 205–214.

Färe, R. S., S. Grosskopf, and C. A. K. Lovell (1994). *Production Frontiers*. Cambridge: Cambridge University Press.

Farrell, M. J. (1957). The measurement of productive efficiency. *Journal of the Royal Statistical Society*, Series A, 120(3), 253–281.

Frisch, R. (1964). *Theory of Production*. Chicago: Rand McNally & Company.

Garey, M. R. and D. S. Johnson (1979). *Computers and Intractability: A Guide to the Theory of NP-Completeness*. W. H. Freeman, San Francisco, CA.

Gautam, N. (2012). *Analysis of Queues: Methods and Applications*. Boca Raton: CRC Press (Taylor & Francis Group).

Gililland, J., J. Konopka, K. Barber, R. Schnabl, and V.A. Ames (1995). Semiconductor manufacturing productivity: Overall equipment effectiveness (OEE) guidelines. Technology transfer 950327443 A-GEN, Revision 1.0. Sematech.

Hackman, S. T. (2008). *Production Economics: Integrating the Microeconomic and Engineering Perspectives*. Heidelberg: Springer-Verlag.

Hanson, D. L. and G. Pledger (1976). Consistency in concave regression. *Annals of Statistics* 4(6): 1038–1050.

Helgeson, W. P. and D. P. Birnie (1961). Assembly line balancing using the ranked positional weight technique. *Journal of Industrial Engineering* 12: 394–398.

Henderson, J. M., and R. E. Quandt (1980). *Microeconomic Theory: A Mathematical Approach*. (3rd ed.). New York: McGraw-Hill.

Hildreth, C. (1954). Point estimates of ordinates of concave functions. *Journal of the American Statistical Association* 49(267): 598–619.

Hopp, W. J. and M. L. Spearman (2004). To pull or not to pull: What is the question? *Manufacturing and Service Operations Management*, 6(2): 133–148.

Jondrow, J., C. A. K. Lovell, I. S. Materov, and P. Schmidt (1982). On the Estimation of Technical Inefficiency in the Stochastic Frontier Production Function Model. *Journal of Econometrics* 19(2-3): 233–238.

Judge, G. G., W. E. Griffiths, R. C. Hill, H. Lutkepohl, and T.-C. Lee (1985). *Introduction to the Theory and Practice of Econometrics*. New York: John Wiley & Sons, Inc.

Koopmans, T. (1951). An analysis of production as an efficient combination of activities. In: Koopmans, T.C. (Ed.), *Activity Analysis of Production and Allocation. Cowles Commission for Research in Economics, Monograph* No. 13. New York: John Wiley & Sons, Inc.

Kuosmanen, T. (2008). Representation theorem for convex nonparametric least squares. *Econometrics Journal* 11: 308–325.

Kuosmanen, T. and A. L. Johnson (2010). Data envelopment analysis as nonparametric least squares regression. *Operations Research* 58(1): 149–160.

Kuosmanen, T. and M. Kortelainen (2012). Stochastic non-smooth envelopment of data: semi-parametric frontier estimation subject to shape constraints. *Journal of Productivity Analysis* 38(1): 11–28.

Liker, J. K. (2004). *The Toyota Way: 14 Management Principles from the World's Greatest Manufacturer*. New York: McGraw-Hill.

Meeusen, W. and J. van den Broeck (1977). Efficiency estimation from Cobb-Douglas production functions with composed error. *International Economic Review* 18(2): 435–444.

Nahmias, S. (2009). *Production and Operations Analysis*. (6th ed.). New York: McGraw-Hill.

Nof, S. Y., W. E. Wilhelm, and H.-J. Warnecke (1997). *Industrial Assembly*. Chapman & Hall, London, UK.

Ohno, T. (1988a). *Toyota Production System: Beyond Large-Scale Production*. Productivity Press Inc., London, UK

Ohno, T. (1988b). *Just-In-Time for Today and Tomorrow*. Productivity Press Inc., London, UK.

Richmond, J. (1974). Estimating the efficiency of production. *International Economic Review* 15: 515–521.

de Ron, A. J. and J. E. Rooda (2005). Equipment effectiveness: OEE revisited. *IEEE Transactions on Semiconductor Manufacturing*, 18(1): 190–196.

Smith, W. (1998). *Time Out: Using Visible Pull Systems to Drive Process Improvements*. New York: John Wiley & Sons.

Standard for Definition and Measurement of Equipment Productivity, Semiconductor Equipment and Material International (SEMI) E79-0200, 2000.

Standard for Definition and Measurement of Equipment Reliability, Availability, and Maintainability, SEMI E10-0701, 2001.

Womack, J. P. and D. T. Jones (2003). *Lean thinking: Banish waste and create wealth in your corporation*. (2nd ed.), New York: Free Press Simon & Schuster.

Peacock, Brian (2019), *Human Systems Integration*, Self-published manuscript, Fernandina Beach, FL.

Peacock, Brian (2021), *Ergonomics Tools and Applications*, Self-published manuscript, Fernandina Beach, FL.

Peacock, Brian (2020), *How Ergonomics Works*, Self-published manuscript, Fernandina Beach, FL.

Rybeck, Charles, Lanny Cornwell, and Phillip Sagan 2017), "Digital Dimension Disruption: A National Security Enterprise Response," *PRISM*, Vol. 7, No. 2, pp. 41–51, 2017.

3 Pervasiveness of AI Technology

AI AND US

In our present digital age, technology is all around us. Artificial intelligence (AI) technology is becoming more and more prevalent in everything we do. The technology is progressing so rapidly to the point of moving from a tool of operational hope to an object of fear, doubt, and apprehension. It behooves our society to find ways to turn technology into constructive assets for whatever we need to do in the specific areas of:

1. Production of physical products (such as rotor casings).
2. Provision of efficient and effective services (such as healthcare services).
3. Generation of results (such as industrial automation with robots).

The tools and techniques of industrial engineering alluded to in the preceding chapters offer tremendous pathways for making AI our positive ally rather than a nemesis: Badiru (1992a, 2022a), Badiru and Cheung (2002a, 2002b), Ibidapo-Obe et al. (1995, 2001), Badiru and Ibidapo-Obe (2008), Badiru et al. (2012), Badiru (2022b), Doran et al (2022).

This chapter covers the spreading pervasiveness of AI in everything we do. From education to manufacturing, AI is spreading rapidly. The premise of this chapter is that the tools and techniques of industrial engineering could help solidify, preserve, and sustain the partnership between humans and technology. The people-technology orientation of industrial engineering means that the role of technology is recognized and appreciated in the context of what humans need and want. Thus, the fear of AI can be allayed if humans come to trust AI more, as a partner rather than as an adversary or competitor. There is nothing artificial in "artificial intelligence." AI is nothing more than data mining infused with computational steroid. What AI is nowadays is what we have always done or aspired to do with our mountains of data. Essentially, What we are doing is data engineering, information engineering, and big Data management, all aided by the present generation of powerful computers that can be networked globally in cloud-based protocols. Some recent and ongoing research interests in academia, where Industrial Engineering (IE) could help develop trustworthy AI, include the following:

- Systems integration matrixing.
- Context-sensitive fusing of networked systems with the goal of discovering bugs in computer software.
- Virtual media programming.
- Community-service infrastructure.
- Enhancement of aerodynamics in aviation.

DOI: 10.1201/9781032659220-3

- Climate prediction.
- Energy transition modes and manners.
- Adaptive curriculum development.
- Sustainability in design.
- Global social engagements.
- Big data design and control.
- Cybersecurity safeguards.
- Information, technology, and communication collaboration.
- Development of advanced materials.
- Autonomous traffic control.
- Financial engineering models.
- Human geography and environment.
- Population growth control.
- Bridging gaps between research and policy.
- War and peace negotiations.
- Visual arts and design.
- Bioengineering.
- Intelligent medical assistants.
- Reconstructive rehabilitation.
- Efficient water access and distribution.
- Agricultural product designs.
- Safety and security silent agents.
- Renewable energy push.
- Critical infrastructure integrity and resilience.
- Miniaturization of intelligent structures.
- Human-embedded sensors.
- Nanofabrication.
- Environmental pollution control.

Everything around us is a system. Embracing a systems-thinking approach is the first order of possibilities for AI. Nothing brings this fact to the front more than the robust discipline of industrial and systems engineering (ISE). The traditional discipline of industrial engineering uses a foundation of systems thinking to facilitate process improvement, efficiency, effectiveness, and productivity in both technical and non-technical areas. The forward-looking approach of industrial engineering makes the embrace of AI natural for industrial engineers. Over the decades, industrial engineering has been at the forefront of the application of new and emerging techniques and technologies. This is evidenced in the early development and utilization of mathematical operations research in the military and the operational leveraging of electronic spreadsheets and computational tools in business and industry in the early days of the emergence of microcomputers (Badiru and Bauer, 1985; Badiru, 1985; Badiru, 1988; Badiru and Whitehouse, 1989; Badiru and Whitehouse 1986; Badiru et al., 1987).

AI is all about fast computation, approximation, estimation, and prediction (see Badiru et al., 1987). In mimicking the human brain, an AI tool must quickly calculate outputs on the order of millions of combinations and permutations. Such a feat is only possible through mathematical foundations, both simple and complex. Every part of the vast expanse of the entire field of mathematics has something to offer to how AI works. One reference is estimation tricks to arrive at a quick answer, similar to using mental math strategies (Balch et al., 1993). Johann Carl Friedrich Gauss (1777–1855), of Gaussian fame, is reputed to have demonstrated amazingly fast math calculation techniques. Gauss was a German mathematician and physicist who made significant contributions to many fields of mathematics and science. Gauss had an exceptional influence in many fields of mathematics and science, and he is ranked among history's most influential mathematicians. He is reported to have impressed his teacher at the age of 10 with his ability to quickly and mentally calculate answers to calculation questions. A story goes that when the teacher asked students to find the sum of the numbers from 1 through 100, young Gauss amazed his teacher and the class by writing the answer immediately. Gauss used a mental math strategy to find the sum. One mental math strategy is compensation. In compensation, you change a problem so it is easy to solve mentally. Then, you adjust to compensate for the change you made. There are many ways to compensate for finding an answer. In many AI techniques, estimation and compensation strategies are used. AI is not just one modeling or computational approach. The diverse applications of AI call for diverse mathematical formulations. This chapter and the chapters that follow present a potpourri of mathematical representations that are directly applicable to AI modeling, programming, and applications. The expectation is that the collection will motivate additional advancements in the area of mathematical foundations for AI in multidisciplinary fields of application.

LEVERAGING THE POTENTIAL OF AI

AI is becoming more pervasive in our modern society. The interest in AI has heightened in recent years, particularly because of the availability of more powerful hardware and software tools. The emerging field of quantum computing bodes well for more robust AI developments in the future. A quantum computer is a computer that uses the quantum states of subatomic particles to store information. The better we can store and retrieve data, the more achievable our AI goals can be.

Quantum computing is the processing of information that is represented by special quantum states. The process taps into quantum phenomena like "superposition" and "entanglement." Thereby creating a fundamentally different way of handling information. When practical quantum computing finally arrives, we will see a dramatic improvement in the operations of computers, laptops, smartphones, and others. Quantum computers can tackle complex problems involving humongous numbers of variables and potential outputs. Simulations and optimization modeling in industrial engineering applications will become faster and more effective. Although quantum computing is not yet where advocates are touting it to be, rapid advancements are happening almost daily. Quantum computing now is where AI used to be some 20 years ago. We now see what AI can do in the present research and development landscape. We expect quantum computing to approach a similar level of growth and practical applications in just a few years more.

Due to the increasing potential of AI in our society, many colleges and universities are starting new formal degree programs in AI. The programs are designed to prepare students for a workforce that is increasingly integrating AI. A key aspect of the program is the emphasis on responsible development, which is a hallmark of the people-oriented profession of industrial engineering, regarding program developments aligned and integrated with the needs of humans. The curriculum includes ethical considerations. Data science is one field where AI is making a noticeable impact.

OLD AND NEW AI APPLICATIONS

What was hypothesized as being within the realm of possibilities several decades ago is now being realized. Although the capabilities have not yet peaked, we can already see applications at home, at work, in leisure pursuits, in sports, in technical undertakings, and so on. Consider the days of physical peepholes on wooden front doors compared to today's digital doorbell cameras. Our society has come a long way in a short period of time. There is a strong wave of interest in AI in every corner of business, industry, government, academia, and military in our society. Sadiku et al. (2022) present several practical AI applications in smart cities, smart electrical grids, healthcare, engineering, education, business, industries, manufacturing, agriculture, vehicle autonomy, social media, and the military. The premise of this book is to focus on selected mathematical framework and foundation for developing sustainable AI tools, techniques, and processes. This is to emphasize that a technology that is based on a mathematical foundation will have a basis for sustainment, very much in the same way that mathematical formulations can stand the test of time.

For years, detractors have derided the prospects for AI as a serious and sustainable tool. But AI can no longer be ignored as more and more realistic AI-based products are developed. AI is not yet where it could be, but it is getting there, albeit slowly.

AI is not something new. It originated as a research curiosity in the 1950s. As far back as 1992, Badiru (1992) highlighted the potential applications of AI in manufacturing, with an explicit emphasis on Expert Systems, the legacy software component of AI. It has taken this long to get to where we are today with AI. Adedeji Badiru developed, directed, and ran a rudimentary AI laboratory at the University of Oklahoma, School of Industrial Engineering, as far back as 1986. It was the first such ambitious university-based technical pursuit in the state of Oklahoma at that time. New AI-based courses were developed, and thesis and dissertation products emanated from the AI efforts of those days. The prediction then was that AI would someday emerge from the laboratory to become a viable asset for the society. We are now beginning to realize that prediction. Some of the scholarly AI products that the authors have been involved in over the past decades are summarized below:

- 1987: Arif, Alaa E. and A. B. Badiru, "An Integrated Expert System with a Fuzzy Linguistic Model for Facilities Layout," *Proceedings of the 6th Oklahoma Symposium on Artificial Intelligence*, Tulsa, Oklahoma, November 11–12, 1992, pp. 185–194.
- 1987: Badiru, A. B., "Applications of Artificial Intelligence in Economic Development," Oklahoma State Legislature, February 1987.
- 1988: Badiru, A. B., Janice M. Mathis, and Bob T. Holloway, "Knowledge Base Design for Law Enforcement," *Proceedings of 10th Annual Conf. on Computers and Industrial Engineering*, Dallas, Texas, March 1988, Computers & Industrial Engineering, Vol. 15, Nos. 1–4, 1988, pp. 78–84.
- 1988: Datar, Neetin N. and A. B. Badiru, "A Prototype Knowledge Based Expert System for Robot Consultancy – ROBCON," in *Proceedings of the*

1988 Oklahoma Symposium on Artificial Intelligence, Norman, Oklahoma, November 1988, pp. 51–68.

- 1988: Joshi, Ajay P.; Neetin N. Datar; and A. B. Badiru, "Knowledge Acquisition and Transfer," in *Proceedings of the 1988 Oklahoma Symposium on Artificial Intelligence*, Norman, Oklahoma, November 1988, pp. 355–378.
- 1990: Badiru, A. B., "Artificial Intelligence Applications in Manufacturing," in *The Automated Factory Handbook: Technology and Management*, Cleland, David I. and Bopaya Bidanda, Editors, TAB Professional and Reference Books, New York, 1990, pp. 496–526.
- 1991: Badiru, A. B., "OKIE-ROOKIE: An Expert System for Industry Relocation Assessment in Oklahoma," *Proceedings of the 5th Oklahoma Symposium on Artificial Intelligence*, Norman, Oklahoma, November 1991.
- 1991: Somasundaram, S. and A. B. Badiru, "An Expert System for External Cylindrical Grinding: Planning, Diagnosing, and Training," *Proceedings of the 5th Oklahoma Symposium on Artificial Intelligence*, Norman, Oklahoma, November 1991, pp. 20–29.
- 1992: Badiru, A. B. (1992), Expert Systems Applications in Engineering and Manufacturing, Prentice-Hall, New Jersey.
- 1992: Badiru, A. B. (1992), Expert Systems Applications in Engineering and Manufacturing, Prentice-Hall, New Jersey.
- 1992: Badiru, A. B.; Le Gruenwald; and Theodore Trafalis, "A New Search Technique for Artificial Intelligence Systems," *Proceedings of the 6th Oklahoma Symposium on Artificial Intelligence*, Tulsa, Oklahoma, November 11–12, 1992, pp. 91–96.
- 1992: Joshi, Shirish; S. Raman; A. B. Badiru; B. L. Foote, "Assembly Interface Determination for Symmetric Rotational Parts," *Proceedings of the 6th Oklahoma Symposium on Artificial Intelligence*, Tulsa, Oklahoma, November 11–12, 1992, pp. 20–29.
- 1993: Badiru, A. B., "Development of a New Search Algorithm for Artificial Intelligence," University of Oklahoma PI Research Investment Program, January 1, 1993 to December, 1993.
- 1993: Badiru, A. B., "The Role of Artificial Intelligence and Expert Systems in New Technologies," in *Management of New Technologies for Global Competitiveness*, Madu, Christian N., Editor, Quorum Books, Greenwood Publishing Company, Westport, Connecticut, 1993, pp. 301–317.
- 1993: Chandler, D., G. Abdelnour, S. Rogers, J. Huang, A. Badiru, J. Cheung, C. Bacon, "Fuzzy Rule-Based System for Early Fault Detection Prediction," *Proceedings of the 7th Oklahoma Symposium on Artificial Intelligence*, Stillwater, Oklahoma, November 18–19, 1993, pp. 62–66.
- 1993: McCauley-Bell, Pamela, "A Fuzzy Linguistic Artificial Intelligence Model for Assessing Risks of Cumulative Trauma Disorders of the Forearm and Hand," Ph.D. Dissertation, University of Oklahoma, 1993.
- 1993: Oliver, Gaugarin E. and A. B. Badiru, "An Expert System Model for Supplier Development Program in a Manufacturing Firm," *Proceedings of the 7th Oklahoma Symposium on Artificial Intelligence*, Stillwater, Oklahoma, November 18–19, 1993, pp. 135–141.

- 1993: Rogers, Steve, "Fuzzy Inferencing Approach to Modeling and Simulation," Ph.D. Dissertation, University of Oklahoma, 1993. Senior Artificial Intelligence Engineering, Seagate Technologies, Oklahoma City.
- 1993: Sieger, David Bruce, "A Methodology for A Real-Time Artificial Intelligence Surveillance System," M.S., School of Industrial Engineering, University of Oklahoma, 1993.
- 1993: Wei, Hong, "Improving Cantor Set Search for Applications in Database and Artificial Intelligence," M.S. (Computer Science), University of Oklahoma, 1993 (co-chair with Dr. Le Gruenwald).
- 1993: Wei, Hong; L. Gruenwald; and A. B. Badiru, "Improving Cantor Set Search for Applications in Database and Artificial Intelligence," *Proceedings of the 7th Oklahoma Symposium on Artificial Intelligence*, Stillwater, Oklahoma, November 18-19, 1993, pp. 250–259.
- 1995: Ibidapo-Obe, O.; A. B. Alonge; and A. B. Badiru, "On Active Controls for a Biped Mechanism," *Applied Mathematics and Computation*, Vol. 69, 1995, pp. 159–183.
- 1996: Milatovic, Milan, "Development of Mode Estimating Technique for Cantor Search of Sorted Data in Artificial Intelligence Systems and Manufacturing Design," M.S. Thesis, School of Industrial Engineering, University of Oklahoma, 1996.
- 1997: Milatovic, M. and A. B. Badiru, "Mode Estimating Procedure for Cantor Searching In Artificial Intelligence Systems," *Proceedings of 1st International Conference on Engineering Design and Automation*, Bangkok, Thailand, March 1997.
- 2001: Ibidapo-Obe, O. Sunday Asaolu, and Adedeji B. Badiru (2001), "Generalized solutions of the pursuit problem in three-dimensional Euclidean space," *Applied Mathematics and Computation*, Vol. 119, No. 3, pp. 35–45.
- 2001: Milatovic, M. and A. B. Badiru, "Control Sequence Generation in Multistage Fuzzy Control Systems for Design Process," *AIEDAM (Artificial Intelligence for Engineering Design, Analysis, and Manufacturing)*, Vol. 15, 2001, pp. 81–87.
- 2002: Badiru, A. B. and John Cheung (2002), *Fuzzy Engineering Expert Systems with Neural Network Applications*, John Wiley & Sons, New York.
- 2002: Badiru, A. B. and John Cheung (2002), *Fuzzy Engineering Expert Systems with Neural Network Applications*, John Wiley & Sons, New York.
- 2004: Badiru, A. B. (2004), "Hybrid NeuroFuzzy Expert System for Crime Investigation," US Department of Justice/Law Enforcement Innovation Center, June 1, 2003—May 31, 2004.
- 2008: Badiru, A. B. and O. Ibidapo-Obe (2008), "On Set-Theoretic Fundamentals of Systems Integration," *African Journal of Computing & ICT (Journal of the IEEE Nigeria Computer Chapter)*, Vol. 1, No. 2, December 2008, pp. 3–25.
- 2008: Olunloyo, V. O. S.; Abayomi Ajofoyinbo; and A. Badiru, "An Alternative Approach for Computing the Union and Intersection of Fuzzy Sets: A Basis for Design of Robust Fuzzy Controller," in *Proceedings of 7th WSEAS International Conference on Artificial Intelligence, Knowledge*

Engineering and Data Bases (AIKED '08), Cambridge, UK, February 23–25, 2008. Note: This paper won best student paper at the conference.
- 2012: Badiru, A. B.; Oye Ibidapo-Obe; and B. J. Ayeni (2012), *Industrial Control Systems: Mathematical and Statistical Models and Techniques*, Taylor & Francis Group, CRC Press, Boca Raton, FL.
- 2022: Badiru, A. B. (2022), *Artificial Intelligence and Digital Systems Engineering*, Taylor & Francis Group, CRC Press, Boca Raton, FL.
- 2022: Badiru, A. B. (2022), *Artificial Intelligence and Digital Systems Engineering*, Taylor & Francis Group, CRC Press, Boca Raton, FL.
- 2022: Doran, Eric, Sharon Bommer, and Adedeji Badiru, "Integration of Human Factors, Cognitive Ergonomics, and Artificial Intelligence in the Human-Machine Interface for Additive Manufacturing," *International Journal of Quality Engineering and Technology, International Journal of Mechatronics and Manufacturing Systems*, Vol. 15, No. 4, 2022, pp. 310–330.

Some of these intellectual products culminated in seminal AI-themed textbooks (Badiru and Cheung, 2002; Badiru, 1992a, b). So, why is AI now bursting into the limelight in light of its long tenure in the dark recesses of many university laboratories? This renewed emergence is due to the emergence of new powerful computational tools and techniques. Below its process orientation, the discipline of industrial engineering had many early influences on the development of AI then and now. Those who know AI inside and out, know how much mathematical programming affects the optimization of AI products. In fact, there is a noticeable application of industrial engineering and operations research techniques of non-linear programming to machine learning and artificial neural networks. Further, people's orientation to industrial engineering paves the way for the consideration of social science and humanities in the practical uses of AI products by people. Nowadays, we are more sensitized to how people use AI products, from a humanistic standpoint rather than from a mechanical standpoint. Similarly, the emergence of more powerful data manipulation techniques created additional inroads into predictive analytics, data envelopment, data quality control, decision science, process efficiency, and path optimization. All of these have foundational linkages to industrial engineering.

An interesting example of a recent AI application is its adoption by FIFA for decision-making in the 2022 Football World Cup in Qatar. AI is deployed to ascertain when the ball has crossed the goal line and also for the detection of "offside" via the usage of sensors and intelligent image/video processing. The virtual assistant referee helps obviate controversies such as England's 1966 cross-bar-related goal in the final, as well as Maradona's (in)famous "hand of God" goal in 1986.

RECAPPING THE IE PLATFORM

So, what does industrial engineering have to do with the surge and resurgence of AI? The answer can be seen in the very definition of industrial engineering, as presented below:

> Industrial engineering is concerned with the design, installation, and improvement of integrated systems of people, materials, information, equipment, and

> energy by drawing upon specialized knowledge and skills in the mathematical, physical, and social sciences, together with the principles and methods of engineering analysis and design to specify, predict, and evaluate the results to be obtained from such systems.

The above definition demonstrates the versatility and diversity of industrial engineering to enmesh with AI. For the purpose of this book, we are focusing on the "mathematical" angle of industrial engineering. The fact that the most fruitful applications of AI can be found in "industry" further solidifies the linkage of the discipline of industrial engineering to the field of AI.

AI is not just one single thing. It is a conglomerate of various elements, involving software, hardware, data platform, policy, procedures, specifications, rules, and people intuition. How we leverage such a multifaceted system to do seemingly intelligent things, typical of how humans think and work, is a matter of systems implementation. This is why the premise of this book centers on a systems methodology. Despite the recent boost in the visibility and hype of AI, it has been around and toyed with for decades. What has brought AI more to the forefront nowadays is the availability and prevalence of high-powered computing tools that have enabled the data-intensive processing required by AI systems. The resurgence of AI has been driven by the following developments:

- Emergence of new computational techniques and more powerful computers.
- Machine learning techniques.
- Autonomous systems.
- New/innovative applications.
- Specialized techniques: intelligent computational search technique using Cantor set sectioning.
- Human-in-the-loop requirements.
- Systems integration aspects.

As long ago as the mid-1980s, the author has led many research and development projects that embedded AI software and hardware into conventional human decision processes. AI has revolutionized and will continue to revolutionize many things we see and use around us. So, we need to pay attention to the emerging developments.

The comprehensive definition of industrial engineering epitomizes what AI is expected to accomplish.

STORY OF THE ORIGIN OF AI

The background of AI has been characterized by controversial opinions and diverse approaches. The controversies have ranged from the basic definition of intelligence to questions about the moral and ethical aspects of pursuing AI. However, despite the unsettled controversies, the technology continues to generate practical results. With increasing efforts in AI research, many of the prevailing arguments are being resolved with proven technical approaches.

"Artificial intelligence" is a controversial name for a technology that promises much potential for improving human productivity. The phrase seems to challenge human pride in being the sole creation capable of possessing real intelligence. All kinds of anecdotal jokes about AI have been offered by casual observers. A speaker once recounted his wife's response when he told her that he was venturing into the new technology of AI. "Thank God, you are finally realizing how dumb I have been saying you were all these years," was alleged to have been the wife's words of encouragement. One whimsical definition of AI refers to it as the "Artificial insemination of knowledge into a machine." Despite the deriding remarks, serious embracers of AI may yet have the last laugh. It is being shown again and again that AI may hold the key to improving operational effectiveness in many areas of applications. Some observers have suggested changing the term "Artificial Intelligence" to a less controversial one such as "Intelligent Applications (IA)." This refers more to the way that computers and software are used innovatively to solve complex decision problems.

Natural intelligence involves the capability of humans to acquire knowledge, reason with the knowledge, and use it to solve problems effectively. It also refers to the ability to develop new knowledge based on existing knowledge. By contrast, AI is defined as the ability of a machine to use simulated knowledge in solving problems. The definition of intelligence had been sought by many ancient philosophers and mathematicians, including Aristotle, Plato, Copernicus, and Galileo. These great philosophers attempted to explain the process of thought and understanding. The real key that started the quest for the simulation of intelligence did not occur, however, until the English philosopher Thomas Hobbes put forth an interesting concept in the 1650s. Hobbes believed that thinking consists of symbolic operations and that everything in life can be represented mathematically. These beliefs directly led to the notion that a machine capable of carrying out mathematical operations on symbols could imitate human thinking. This is the basic driving force behind the AI effort. For that reason, Hobbes is sometimes referred to as the grandfather of AI.

While the term "Artificial Intelligence" was coined by John McCarthy relatively recently (1956), the idea had been considered centuries before. As early as 1637, Rene Descartes was conceptually exploring the ability of a machine to have intelligence when he said:

For we can well imagine a machine so made that it utters words and even, in a few cases, words pertaining specifically to some actions that affect it physically. However, no such machine could ever arrange its words in various different ways so as to respond to the sense of whatever is said in its presence—as even the dullest people can do.

Descartes believed that the mind and the physical world are on parallel planes that cannot be equated. They are of different substances following entirely different rules and can, thus, not be successfully compared. The physical world (i.e., machines) cannot imitate the mind because there is no common reference point.

Hobbes proposed the idea that thinking could be reduced to mathematical operations. On the other hand, Descartes had the insight into functions that machines might someday be able to perform. However, he had reservations about the concept that thinking could simply be a mathematical process.

The 1800s was an era that saw some advancement in the conceptualization of the computer. Charles Babbage, a British mathematician, laid the foundation for the construction of the computer, a machine defined as being capable of performing mathematical computations. In 1833, Babbage introduced an analytical engine. This computational machine incorporated two unprecedented ideas that were to become crucial elements in the modern computer. First, it had operations that were fully programmable, and second, the engine could contain conditional branches. Without these two abilities, the power of today's computers would be inconceivable. Babbage was never able to realize his dream of building the analytic engine due to a lack of financial support. However, his dream was revived through the efforts of later researchers. Babbage's basic concepts could be observed in the way that most computers operate today.

Another British mathematician, George Boole, worked on issues that were to become equally important. Boole formulated the "Laws of Thought" that set up rules of logic for representing thought. The rules contained only two-valued variables. By this, any variable in a logical operation could be in one of only two states: yes or no, true or false, all or nothing, 0 or 1, on or off, and so on. This was the birth of digital logic, a key component of the AI effort.

In the early 1900s, Alfred North Whitehead and Bertrand Russell extended Boole's logic to include mathematical operations. This not only led to the formulation of digital computers but also made possible one of the first ties between computers and thought process.

However, there was still a lack of an acceptable way to construct such a computer. In 1938, Claude Shannon published "A Symbolic Analysis of Relay and Switching Circuits." This work demonstrated that Boolean logic consisting of only two-variable states (e.g. on-off switching of circuits) can be used to perform logic operations. Based on this premise, the Electronic Numerical Integrator and Computer (ENIAC) was built in 1946 at the University of Pennsylvania. The ENIAC was a large-scale fully operational electronic computer that signaled the beginning of the first generation of computers. It could perform calculations 1000 times faster than its electromechanical predecessors. It weighed 30 tons, stood two stories high, and occupied 1500 square feet of floor space. Unlike today's computers that operate in binary codes (0s and 1s), the ENIAC operated in decimal (0, 1, 2, …, 9), and it required 10 vacuum tubes to represent one decimal digit. With over 18,000 vacuum tubes, the ENIAC needed a great amount of electrical power; so much that it was said that it dimmed the lights in Philadelphia whenever it operated.

HUMAN INTELLIGENCE VERSUS AI

Two of the leading mathematicians and computer enthusiasts during the 1900–1950 time frame were Alan Turing and John Von Neumann. In 1945, Von Neumann insisted that computers should not be built as glorified adding machines, with all their operations specified in advance. Rather, he suggested, computers should be built as general-purpose logic machines capable of executing a wide variety of programs. Such machines, Von Neumann proclaimed, would be highly flexible and capable of being readily shifted from one task to another. They could react intelligently to the results of their calculations, could choose among alternatives, and could even play checkers or chess. This represented something unheard of at that time: a machine with built-in intelligence, able to operate on internal instructions.

Prior to Von Neumann's concept, even the most complex mechanical devices had always been controlled from the outside, for example, by setting dials and knobs. Von Neumann did not invent the computer, but what he introduced was equally significant: computing by use of computer programs, the way it is done today. His work paved the way for what would later be called artificial intelligence in computers.

Alan Turing also made major contributions to the conceptualization of a machine that can be universally used for all problems based only on variable instructions fed into it. Turing's universal machine concept, along with Von Neumann's concept of a storage area containing multiple instructions that can be accessed in any sequence, solidified the ideas needed to develop the programmable computer. Thus, a machine was developed that could perform logical operations and could do them in varying orders by changing the set of instructions that were executed.

Due to the fact that operational machines were now being realized, questions about the "intelligence" of the machines began to surface. Turing's other contribution to the world of AI came in the area of defining what constitutes intelligence. In 1950, he designed the Turing test for determining the intelligence of a system. The test utilized the conversational interaction between three players to try and verify computer intelligence.

The test is conducted by having a person (the interrogator) in a room that contains only a computer terminal. In an adjoining room, hidden from view, a man (Person A) and a woman (Person B) are located with another computer terminal. The interrogator communicates with the couple in the other room by typing questions on the keyboard. The questions appear on the couple's computer screen, and they respond by typing on their own keyboard. The interrogator can direct questions to either Person A or Person B but without knowing which is the man and which is the woman.

The purpose of the test is to distinguish between the man and the woman merely by analyzing their responses. In the test, only one of the people is obligated to give truthful responses. The other person deliberately attempts to fool and confuse the interrogator by giving responses that may lead to an incorrect guess. The second stage of the test is to substitute a computer for one of the two persons in the other room. Now, the human is obligated to give truthful responses to the interrogator while the computer tries to fool the interrogator into thinking that it is human. Turing's contention is that if the interrogator's success rate in the human/computer version of the game is not better than his success rate in the man/woman version, then the computer can be said to be "thinking." That is, the computer possesses "intelligence." Turing's test has served as a classical example for AI proponents for many years.

By 1952, computer hardware had advanced far enough that actual experiments in writing programs to imitate thought processes could be conducted. The team of Herbert Simon, Allen Newell, and Cliff Shaw organized to conduct such an experiment. They set out to establish what kinds of problems a computer could solve with the right programming. Proving theorems in symbolic logic such as those set forth by Whitehead and Russell in the early 1900s fit the concept of what they felt an intelligent computer should be able to handle.

It quickly became apparent that there was a need for a new, higher-level computer language than was currently available. First, they needed a language that was more user-friendly and could take program instructions that are easily understood by a human programmer and automatically convert them into machine language that could be understood by the computer. Second, they needed a programming language that changed the way in which computer memory was allocated. All previous languages would pre-assign memory at the start of a program. The team found that the type of programs they were writing would require large amounts of memory and would function unpredictably.

To solve the problem, they developed a list-processing language. This type of language would label each area of memory and then maintain a list of all available memory. As memory became available, it would update the list, and when more memory was needed, it would allocate the amount necessary. This type of programming also allowed the programmer to be able to structure his or her data so that any information that was to be used for a particular problem could be easily accessed.

The end result of their effort was a program called Logic Theorist. This program had rules consisting of axioms already proved. When it was given a new logical expression, it would search through all of the possible operations in an effort to discover proof of the new expression. Instead of using a brute force search method, they pioneered the use of heuristics in the search method.

The Logic Theorist that they developed in 1955 was capable of solving 38 of 52 theorems that Whitehead and Russell had devised. It was not only capable of the proofs but did them very quickly. What took a Logic Theorist a matter of minutes to prove would have taken years to do if it had been done by simple brute force on a computer. By comparing the steps which it went through to arrive at a proof to those that human subjects went through, it was also found that it had a remarkable imitation of the human thought process.

NATURAL LANGUAGE IMPLICATIONS

Despite the various successful experiments, many observers still believe that AI does not have much potential for practical applications. There is a popular joke in the AI community that points out the deficiency of AI in natural language applications. It is said that a computer was asked to translate the following English statement into Russian and back to English: "*The spirit is willing but the flesh is weak.*" The reverse translation from Russian to English yielded: "*The vodka is good but the meat is rotten.*"

From my own author perspective, AI systems are not capable of thinking in the human sense. They are great at mimicking based on the massive amounts of data structures and linkages available. For example, consider the following natural language interpretations of the following ordinary statements:

"NO SALT IS SODIUM FREE"

A human being can quickly infer the correct interpretation and meaning based on the prevailing context of the conversation. However, an "intelligent" machine may see the same statement in different ways, as enumerated below:

> **"No (salt) is sodium free,"** which negates the property of the object, salt. This means that there is no type of salt that is sodium free. In other words, all salts contain sodium.

Alternately, the statement can be seen as follows:

> **"(No-salt) is sodium free,"** which is a popular advertisement slogan for the commercial kitchen ingredient named (No-salt). In this case, the interpretation is that this product, named No-salt does not contain sodium.

Here is another one:

> "No news is good news."

This is a common saying that humans can easily understand regardless of the context. In AI reasoning, it could be subject to the following interpretations:

> **"(No news) is good news,"** which agrees with the normal understanding that the state of having no news implies the absence of bad news, which is good (i.e. desirable). In this case, (No-news), as a compound word, is the object.

Or, an AI system could see it as:

> **"No (news) is good news,"** which is a contradiction of the normal interpretation. In this case, the AI system could interpret it as a case where all pieces of news are bad (i.e., not good). This implies that the object is the (news).

Here is another one from the political arena:

> "The British parliament wants no deal off the table."

The AI interpretations could see the objects as follows:

(No deal), as a condition of negotiation, is off the table.

Alternately, it could be seen, in the negation sense, as all deals are acceptable to be on the table.

Pattern recognition is another interesting example that distinguishes human intelligence from machine intelligence. For example, when I park my vehicle in a large shopping center parking lot, with many similarly colored and shaped vehicles, I can always identify my vehicle from a far distance by simply seeing a tiny segment of the body of the vehicle. This could be by seeing half of the headlight jotting out from among several vehicles. It could be by seeing a portion of the tail light. It could even be by seeing the luggage carriage on top of the vehicle barely visible above other vehicles around mine. For an AI system to use pattern recognition to correctly identify my vehicle, it would have to use a tremendous amount of data collection, data manipulation, interpolation, extrapolation, and other complex mathematical algorithms to consider the options to provide a probable match. Considering examples such as the above, it can be seen that human intelligence and natural perception still trump machine's simulated intelligence. Despite of this deficiency, machine intelligence, under the banner of AI, can be useful to supplement human intelligence to arrive at a more efficient and effective decision process. Consequently, AI is a useful and desirable ally in human operations. This belief was what drove the early efforts in defining and advancing the science, technology, engineering, and mathematics foundations for AI.

THE FIRST AI CONFERENCE

The summer of 1956 signified the first attempt to establish the field of machine intelligence into an organized effort. The Dartmouth Summer Conference, organized by John McCarthy, Marvin Minsky, Nathaniel Rochester, and Claude Shannon, brought together people whose work and interests formally founded the field of AI. The conference, held at Dartmouth College in New Hampshire, was funded by a grant from the Rockefeller Foundation. It was at that conference that John McCarthy coined the term "artificial intelligence." It was the same John McCarthy who developed the LISP programming language that has become a standard tool for AI development. In attendance at the meeting, in addition to the organizers, were Herbert Simon, Allen Newell, Arthur Samuel, Trenchard More, Oliver Selfridge, and Ray Solomon.

The Logic Theorist, developed by Newell, Shaw, and Simon, was discussed at the conference. The system, considered the first AI program, used heuristic search to solve mathematical problems in *Principia Mathematica*, written by Whitehead and Russell). Newell and Simon were far ahead of others in actually implementing AI ideas with their Logic Theorist. The Dartmouth meeting served mostly as an avenue for the exchange of information and, more importantly, as a turning point in the main emphasis of work in the AI endeavor. Instead of concentrating so much on the hardware to imitate intelligence, the meeting set the course for examining the structure of the data being processed by computers, the use of computers to process symbols, the need for new languages, and the role of computers for testing theories.

EVOLUTION OF SMART PROGRAMS

The next major step in software technology came from Newell, Shaw, and Simon in 1959. The program they introduced was called General Problem Solver (GPS). GPS was intended to be a program that could solve many types of problems. It was capable of solving theorems, playing chess, or doing various complex puzzles. GPS was a significant step forward in AI. It incorporates several new ideas to facilitate problem-solving. The nucleus of the system was the use of means-end analysis. Means-end analysis involves comparing a present state with a goal state. The difference between the two states is determined and a search is done to find a method to reduce this difference. This process is continued until there is no difference between the current state and the goal state.

In order to further improve the search, GPS contained two other features. The first is that, if while trying to reduce the deviation from the goal state, it finds that it has actually complicated the search process, it was capable of backtracking to an earlier state and exploring alternate solution paths. The second is that it was capable of defining sub-goal states that, if satisfied, would permit the solution process to continue. In formulating GPS, Newell and Simon had done extensive work studying human subjects and the way they solved problems. They felt that GPS did a good job of imitating the human subjects. They commented on the effort by saying (Newell and Simon 1961):

The fragmentary evidence we have obtained to date encourages us to think that the General Problem Solver provides a rather good first approximation to an information processing theory of certain kinds of thinking and problem-solving behavior. The processes of "thinking" can no longer be regarded as completely mysterious.

GPS was not without critics. One of the criticisms was that the only way the program obtained any information was to get it from human input. The way and order in which the problems were presented was controlled by humans, thus the program was doing only what it was told to do. Newell and Simon argued that the fact that the program was not just repeating steps and sequences but was actually applying rules to solve problems it had not previously encountered is indicative of intelligent behavior.

There were other criticisms also. Humans are able to devise new shortcuts and improvise. GPS would always go down the same path to solve the same problem, making the same mistakes as before. It could not learn. Another problem was that GPS was good when given a certain area or a specific search space to solve. The problem with this limitation was that in the solution of problems, it was difficult to determine what search space to use. Sometimes, solving the problem is trivial compared to finding the search space. The problems posed to GPS were all of a specific nature. They were all puzzles or logical challenges: problems that could easily be expressed in symbolic form and operated on in a pseudo-mathematical approach. There are many problems that humans face that are not so easily expressed in symbolic form.

Also during the year 1959, John McCarthy came out with a tool that was to greatly improve the ability of researchers to develop AI programs. He developed a new computer programming language called list processing (LISP). It was to become one of the most widely used languages in the field.

LISP is distinctive in two areas: memory organization and control structure. The memory organization is done in a tree fashion with interconnections between memory groups. Thus, it permits a programmer to keep track of complex structural relationships. The other distinction is the way the control of the program is done. Instead of working from the prerequisites to a goal, it starts with the goal and works backward to determine what prerequisites are required to achieve the goal.

In 1960, Frank Rosenblatt did some work in the area of pattern recognition. He introduced a device called PERCEPTRON that was supposed to be capable of recognizing letters and other patterns. It consisted of a grid of 400 photocells connected with wires to a response unit that would produce a signal only if the light coming off the subject to be recognized crossed a certain threshold.

During the latter part of the 1960s, there were two efforts in another area of simulating human reasoning. Kenneth Colby at Stanford University and Joseph Weizenbaum at MIT wrote separate programs that were capable of interacting in a two-way conversation. Weizenbaum's program was called ELIZA. The programs were able to sustain very realistic conversations by using very clever techniques.

For example, ELIZA used a pattern matching method that would scan for keywords like "I," "you," "like," and so on. If one of these words was found, it would execute rules associated with it. If there was no match found, the program would respond with a request for more information or with some noncommittal response.

It was also during the 1960s that Marvin Minsky and his students at MIT made significant contributions toward the progress of AI. One student, T. G. Evans, wrote a program that could perform visual analogies. The program was shown two figures that had some relationship to each other and was then asked to find another set of figures from a set that matched the same relationship. The input to the computer was not done by a visual sensor (like the one worked on by Rosenblatt), but instead the figures were described to the system.

In 1968, another student of Minsky's, Daniel Bobrow, introduced a linguistic problem solver called STUDENT. It was designed to solve problems that were presented to it in a word problem format. The key to the program was the assumption that every sentence was an equation. It would take certain words and turn them into mathematical operations. For example, it would convert "is" into "=" and "per" into "/".

Even though STUDENT responded very much the same way that a real student would, there was a major difference in depth of understanding. While the program was capable of calculating the time two trains would collide given the starting points and speeds of both, it had no real understanding or even cared what a "train" or "time" was. Expressions like "per chance" and "this is it" could mean totally different things than what the program would assume. A human student would be able to discern the intended meaning from the context in which the terms were used.

In an attempt to answer the criticisms about understanding, another student at MIT, Terry Winograd, developed a significant program named SHRDLU. In setting up this program, he utilized what was referred to as a micro-world or blocks-world. This limited the scope of the world that the program had to try to understand. The program communicated in what appeared to be natural language.

The operation of SHRDLU consisted of a set of blocks of varying shapes (cubes, pyramids, etc.), sizes, and colors. These blocks were all set on an imaginary table. Upon request, SHRDLU would rearrange the blocks to any requested configuration. The program was capable of knowing when a request was unclear or impossible. For instance, if it was requested to put a block on top of the pyramid, it would request that the user specify more clearly what block and what pyramid. It could also recognize that the block would not sit on top of the pyramid.

Two other approaches that the program took that were new to programs were the ability to make assumptions and the ability to learn. If asked to pick up a larger block, it would assume a larger block than the one it was currently working on. If asked to build a figure that it did not know, it would ask for an explanation of what it was and, thereafter, it would recognize the object. One major sophistication that SHRDLU added to the science of AI programming was its use of a series of expert modules or specialists. There was one segment of the program that specialized in segmenting sentences into meaningful word groups, a sentence specialist to determine the relationship between nouns and verbs, and a scenario specialist that understood how individual scenes related to one another. This sophistication added much enhancement to the method in which instructions were analyzed.

As sophisticated as SHRDLU was at that time, it did not escape criticism. Other scholars were quick to point out its deficiencies. One of the shortcomings was that SHRDLU only responded to requests; it could not initiate conversations. It also had no sense of conversational flow. It would jump from performing one type of task to a totally different one if so requested. While SHRDLU had an understanding of the tasks it was to perform and the physical world in which it operated, it still could not understand very abstract concepts.

EMERGING OF AI

The various attempts to formally define the use of machines to simulate human intelligence led to the development of several branches of AI. Current sub-specialties of AI include:

(1) *Natural language processing*: This deals with various areas of research such as database inquiry systems, story understanders, automatic text indexing, grammar, and style analysis of text, automatic text generation, machine translation, speech analysis, and speech synthesis.
(2) *Computer vision*: This deals with research efforts involving scene analysis, image understanding, and motion derivation.
(3) *Robotics*: This involves the control of effectors on robots to manipulate or grasp objects, locomotion of independent machines, and use of sensory input to guide actions.
(4) *Problem solving and planning*: This involves applications such as refinement of high-level goals into lower-level ones, determination of actions needed to achieve goals, revision of plans based on intermediate results, and focused search of important goals.
(5) *Learning*: This area of AI deals with research into various forms of learning, including rote learning, learning through advice, learning by example, learning by task performance, and learning by following concepts.
(6) *Expert systems*: This deals with the processing of knowledge as opposed to the processing of data. It involves the development of computer software to solve complex decision problems.
(7) *ChatGPT*: With the software training capabilities of artificial neural networks, a new AI has emerged, named AI Chatbot, the most famous version of which is ChatGPT. ChatGPT is a free and easy-to-use AI app that can help with writing, learning, brainstorming, strategizing, and so on. Users can type or talk to ChatGPT and get reasonable answers. Writing and editing efficiency, effectiveness, and productivity improvements have been documented for ChatGPT. One commencement speaker at a major university in 2023 wrote his entire talk with the help of ChatGPT. The point here is that ChatGPT is spreading like wildfire. Educators are split on its acceptance. Is it cheating or assistance? From an industrial engineering perspective, any tool that can aid efficiency, effectiveness, and productivity should be embraced and accepted as an operational partner.

NEURAL NETWORKS

Neural networks, sometimes called connectionist systems, represent networks of simple processing elements or nodes capable of processing information in response to external inputs. Neural networks were originally presented as being models of the human nervous system. Just after World War II, scientists found out that the physiology of the brain was similar to the electronic processing mode used by computers. In both cases, large amounts of data are manipulated. In the case of computers, the elementary unit of processing is the bit, which is in either an "on" or "off" state. In the case of the brain, *neurons* perform the basic data processing. Neurons are tiny cells that follow a binary principle of being either in a state of firing (on) or not firing (off). When a neuron is on, it fires a signal to other neurons across a network of synapses.

In the late 1940s, Donald Hebb, a researcher, hypothesized that biological memory results when two neurons are active simultaneously. The synaptic connection of synchronous neurons is reinforced and given preference over connections made by neurons that are not active simultaneously. The level of preference is measured as a weighted value. Pattern recognition, a major strength of human intelligence, is based on the weighted strengths of the reinforced connections between various pairs of simultaneously active neurons.

The idea presented by Hebb was to develop a computer model based on the way in which neurons form connections in the human brain. But the idea was considered to be preposterous at that time since the human brain contains 100 billion neurons and each neuron is connected to 10,000 others by a synapse. Even by today's computing capability, it is still difficult to duplicate the activities of neurons. In 1969, Marvin Minsky and Seymour Pappert wrote the book titled *Perceptrons*, in which they criticized existing neural network research as being worthless. It has been claimed that the pessimistic views presented by the book discouraged further funding for neural network research for several years. Funding was, instead, diverted to further research of expert systems, which Minsky and Pappert favored. It is only recently that neural networks are beginning to make a strong comeback.

Because neural networks are modeled after the operations of the brain, they hold considerable promise as building blocks for achieving the ultimate aim of AI. The present generation of neural networks uses artificial neurons. Each neuron is connected to at least one other neuron in a synapse-like fashion. The networks are based on some form of learning model. Neural networks learn by evaluating changes in input. Learning can be either supervised or unsupervised. In supervised learning, each response is guided by given parameters. The computer is instructed to compare any inputs to ideal responses, and any discrepancy between the new inputs and ideal responses is recorded. The system then uses this data bank to guess how much the newly gathered data is similar to or different from the ideal responses. That is, how closely the pattern matches. Supervised learning networks are now commercially used for control systems and for handwriting and speech recognition.

In unsupervised learning, input is evaluated independently and stored as patterns. The system evaluates a range of patterns and identifies similarities and dissimilarities among them. However, the system cannot derive any meaning from the information without human assignment of values to the patterns. Comparisons are relative to other results, rather than to an ideal result. Unsupervised learning networks are used to discover patterns where a particular outcome is not known in advance, such as in physics research and the analysis of financial data. Several commercial neural network products are now available. An example is NeuroShell from Ward Systems Group. The software is expensive, but it is relatively easy to use. It interfaces well with other software such as Lotus 1-2-3 and dBASE, as well as with C, Pascal, FORTRAN, and BASIC programming languages.

Despite the proven potential of neural networks, they drastically oversimplify the operations of the brain. The existing systems can undertake only elementary pattern-recognition tasks and are weak at deductive reasoning, math calculations, and other computations that are easily handled by conventional computer processing. The difficulty in achieving the promise of neural networks lies in our limited understanding of how the human brain functions. Undoubtedly, to accurately model the brain, we must know more about it. However, a complete knowledge of the brain is still many years away.

Google's ANN-based "DeepMind" is arguably one of the strongest AI engines available at the moment; it has been used to tackle numerous problems in Mathematics, Chess, etc.

APPLICATIONS OF EXPERT SYSTEMS

In the late 1960s to early 1970s, a special branch of AI began to emerge. The branch, known as expert systems, has grown dramatically in the past few years, and it represents the most successful demonstration of the capabilities of AI. Expert systems are the first truly commercial application of work done in the AI field and as such have received considerable publicity. Due to the potential benefits, there is currently a major concentration in the research and development of expert systems compared to other efforts in AI.

Not driven by the desire to develop general problem-solving techniques that had characterized AI before, expert systems address problems that are focused. When Edward Feigenbaum developed the first successful expert system, DENDRAL, he had a specific type of problem that he wanted to be able to solve. The problem involved determining which organic compound was being analyzed in a mass spectrograph. The program was intended to simulate the work that an expert chemist would do in analyzing the data. This led to the term expert system.

The period of time from 1970 to 1980 saw the introduction of numerous expert systems to handle several functions from diagnosing diseases to analyzing geological exploration information. Of course, expert systems have not escaped the critics. Given the nature of the system, critics argue that it does not fit the true structure of AI. Because of the use of only specific knowledge and the ability to solve only specific problems, some critics are apprehensive about referring to an expert system as being intelligent. Proponents argue that if the system produces the desired results, it is of little concern whether it is intelligent or not.

A controversy of interest surfaced in 1972 with a book published by Hubert Dreyfus called *What Computers Can't Do: A Critique of Artificial Reason.* Views similar to those contained in the book were presented in 1976 by Joseph Weizenbaum. The issues that both authors raised touched on some of the basic questions that prevailed way back in the days of Descartes. One of Weizenbaum's reservations concerned what should ethiçally and morally be handed over to machines. He maintained that the path that AI was pursuing was headed in a dangerous direction. There are some aspects of human experience, such as love and morality, that could not adequately be imitated by machines.

While the debates were going on over how much AI could do, the work on getting AI to do more continued. In 1972, Roger Shrank introduced the notion of script: the set of familiar events that can be expected from an often encountered setting. This enables a program to quickly assimilate facts. In 1975, Marvin Minsky presented the idea of frames. Even though both concepts did not drastically advance the theory of AI, they did help expedite research in the field.

In 1979, Minsky suggested a method that could lead to a better simulation of intelligence. He presented the "society of minds" view, in which the execution of knowledge is performed by several programs working in conjunction simultaneously. This concept helped to encourage interesting developments such as present-day parallel processing.

As time proceeded through the 1980s, AI gained significant exposure and interest. AI, once a phrase restricted to the domain of esoteric research, has now become a practical tool for solving real problems. While AI is enjoying its most prosperous period, it is still plagued with disagreements and criticisms. The emergence of commercial expert systems on the market has created both enthusiasm and skepticism. There is no doubt that more research and successful application developments will help prove the potential of expert systems. It should be recalled that new technologies sometimes fail to convince all initial observers. IBM (International Business Machine), which later became a giant in the personal computer business, hesitated for several years before getting into the market because the company never thought that those little boxes called personal computers would ever have any significant impact on the society. How wrong they were!

The effort in AI is a worthwhile endeavor as long as it increases the understanding that we have of intelligence and as long as it enables us to do things that we previously could not do. Due to the discoveries made in AI research, computers are now capable of things that were once beyond imagination.

Embedded Expert Systems: More expert systems are beginning to show up, not as stand-alone systems but as software applications in large software systems. This trend is bound to continue as systems integration takes hold in many software applications. Many conventional commercial packages such as statistical analysis systems, data management systems, information management systems, project management systems, and data analysis systems now contain embedded heuristics that constitute expert systems components of the packages. Even some computer operating systems now contain embedded expert systems designed to provide real-time systems monitoring and troubleshooting. With the success of embedded expert systems, the long-awaited payoffs from the technology are now beginning to be realized.

Because the technology behind expert systems has changed little over the past decade, the issue is not whether the technology is useful but how to implement it. This is why the integrated approach of this book is very useful. This book focuses not only on the technology of expert systems but also on how to implement and manage the technology. Combining neural network technology with expert systems, for example, will become more prevalent. In combination, the neural network might be implemented as a tool for scanning and selecting data, while the expert system would evaluate the data and present recommendations.

While AI technology is good and amenable to organizational objectives, we must temper it with human intelligence. The best hybrid is when machine intelligence is integrated with human intelligence. The human can handle the intuition part while the machine, as AI, handles the data-intensive and number-crunching parts.

Over the past several years, expert systems have proven their potential for solving important problems in engineering and manufacturing environments. Expert systems are helping major companies to diagnose processes in real time, schedule operations, troubleshoot equipment, maintain machinery, and design service and production facilities. With the implementation of expert systems in industrial environments, companies are finding that real-world problems are best solved by an integrated strategy involving the management of personnel, software, and hardware systems.

Solutions to most engineering and manufacturing problems involve not only heuristics but also mathematical calculations, large data manipulations, statistical analysis, real-time information management, system optimization, and man-machine interfaces. These issues and other related topics are addressed in detail in this book. In addition to the basic concepts of expert systems, guidelines are presented on various items ranging from problem selection; data analysis; knowledge acquisition; and system development to verification, validation, integration, implementation, and maintenance.

Can AI systems and products live up to the hype?

In general, the expectations for all products, systems, and processes of AI include effectiveness, efficiency, ease of use, elegance, safety, security, sustainability, and satisfaction (Doran et al 2022, Badiru and Bauer 1985). A systems view can, indeed, bring us closer to realizing these expectations. In terms of a practical and readily seen example of AI, look no further than your mobile phone. The present generation of smartphones is readily a common example of an AI system. So, AI is already all around us on a daily basis. Figure 3.1 illustrates a comprehensive view of IE and DE (Digital Engineering) for AI. It can be seen that every element within the definition of industrial engineering fits within the wide scope of AI. Liebowitz (2021) presents a wide coverage of digital platforms using data analytics to enable AI applications in diverse fields. Some of the topics include autonomy, machine intelligence, managerial decision-making, deep learning, natural language processing, smart materials, predicting, and cognitive neuroimaging. Lohr (2022) describes how one entrepreneur, David Ferrucci, is fusing common sense into AI applications. It is recalled that common sense applications and implications are among the bastions of industrial engineering.

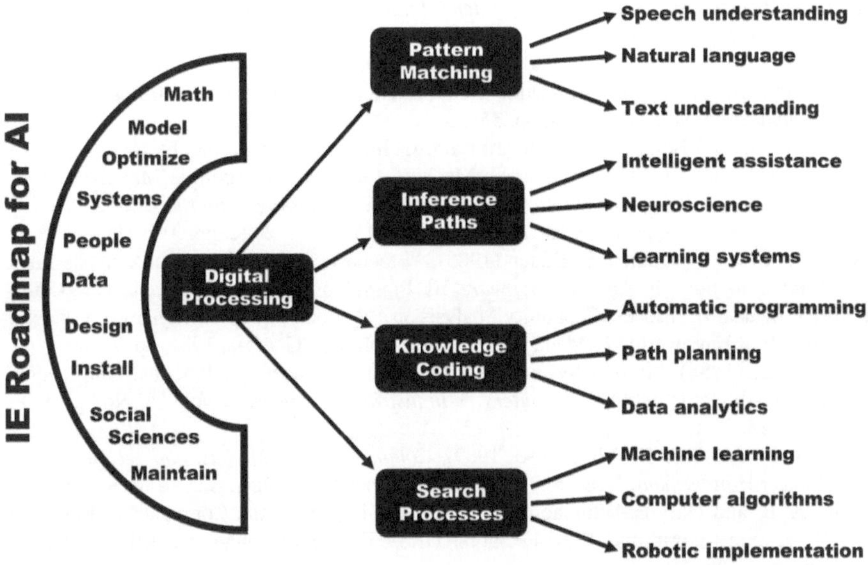

FIGURE 3.1 Industrial Engineering Roadmap for AI.

REFERENCES

Badiru, A. B. and John Cheung (2002a), *Fuzzy Engineering Expert Systems with Neural Network Applications*, John Wiley & Sons, New York.

Badiru, A. B. (1992a), *Expert Systems Applications in Engineering and Manufacturing*, Prentice-Hall, New Jersey.

Badiru, A. B. (2022a), *AI and Digital Systems Engineering*, Taylor & Francis Group, CRC Press, Boca Raton, FL.

Liebowitz, Jay editor (2021), *Data Analytics and AI*, CRC Press/Taylor & Francis Group, Boca Raton, FL.

Lohr, Steve (2022), "One Man's Dream of Fusing AI, Common Sense," *The New York Times*, Technology Section, November 13, 2022.

Sadiku, Matthew N. O., S. M. Musa, and S. R. Nelatury (2022), *Applications of AI*, Gotham Books, Sheridan, WY.

Badiru, A. B. and John Cheung (2002b), *Fuzzy Engineering Expert Systems with Neural Network Applications*, John Wiley & Sons, New York.

Badiru, A. B. (1992b), *Expert Systems Applications in Engineering and Manufacturing*, Prentice-Hall, New Jersey

Ibidapo-Obe, O., A. B. Alonge, and A. B. Badiru, "On Active Controls for a Biped Mechanism," *Applied Mathematics and Computation*, Vol. 69, 1995, pp. 159–183.

Badiru, A. B. and O. Ibidapo-Obe (2008), "On Set-Theoretic Fundamentals of Systems Integration," *African Journal Of Computing & ICT (Journal of the IEEE Nigeria Computer Chapter)*, Vol. 1, No. 2, December 2008, pp. 3–25.

Badiru, A. B., Oye Ibidapo-Obe, and B. J. Ayeni (2012), *Industrial Control Systems: Mathematical and Statistical Models and Techniques*, Taylor & Francis Group, CRC Press, Boca Raton, FL.

Balch, Kay et al (1993), *Mathematics: Applications and Connections*, MacMillan, New York, NY.

Badiru, A. B. (2022b), *AI and Digital Systems Engineering*, Taylor & Francis Group, CRC Press, Boca Raton, FL.

Ibidapo-Obe, O., Sunday Asaolu, and Adedeji B. Badiru (2001), "Generalized Solutions of the Pursuit Problem in Three-Dimensional Euclidean Space," *Applied Mathematics and Computation*, Vol. 119, No. 3, pp. 35–45.

Doran, Eric, Sharon Bommer, and Adedeji Badiru, "Integration of Human Factors, Cognitive Ergonomics, and AI in the Human-Machine Interface for Additive Manufacturing," *International Journal of Quality Engineering and Technology, International Journal of Mechatronics and Manufacturing Systems*, Vol. 15, No. 4, 2022, pp. 310–330.

Badiru, A. B. and Christian S. Bauer (1985), "Stochastic Rate of Return Analysis on a Microcomputer," in *Softcover Software*, IIE Press, Norcross, GA, 1985, pp. 127–135.

Badiru, A. B. (1985), "Process Capability Analysis on a Microcomputer," in *Softcover Software*, Industrial Engineering & Management Press, Norcross, Georgia, 1985, pp. 7–14.

Badiru, A. B. (1988), "Expert Systems and Industrial Engineers: A Practical Guide for a Successful Partnership," *Computers & Industrial Engineering*, Vol. 14, No. 1, 1988, pp. 1–13.

Badiru, A. B. and Gary E. Whitehouse (1989), *Computer Tools, Models, and Techniques for Project Management*, TAB Professional & Reference Books, Inc., Blue Ridge Summit, PA

Badiru, A. B. and Gary E. Whitehouse (1986), "The Impact of the Computer on Resource Allocation Algorithms," Presented at the *ORSA/TIMS Fall Conference*, Miami, Florida, October 1986.

Badiru, A. B., J. Mathis, and B. Holloway, (1987) "AREST Expert System To Aid Crime Investigations," *AI Interactions, Texas Instruments Data Systems*, Vol. 3, No. 3, November 1987, pp. 5–6.

4 AI Control of Complex Systems*

By Md Jubair, Hossain Md Sabbir, Ngulube Lazarous, Hantosh Ali Hasan Abdulkarem, Shimul Paul, Muhammad Sabeeh Khan, Mohammed Arfan, and Abdullah al Rakib

* Adapted by Adedeji B. Badiru.
Adapted under Open Access (CC BY license) from: Jubair, M. D. et al. (2024), "A Comprehensive Review on Intelligence Control for Complex System," *International Journal of Advanced Engineering Research and Science (IJAERS)*, Vol. 11, Issue 1, pp. 44–49, Jan, 2024, https://ijaers.com/. DOI: https://doi.org/10.22161/ijaers.11.7.

INTRODUCTION

Control system intellectualization issues are observed (Jubair et al. 2024) The need for the intellectualization of a diverse variety of systems and control approaches is supported. The hierarchy of intellectual control levels is examined, and various artificial intelligence (AI) methods are compared. Intelligence control for complex systems involves using advanced algorithms and techniques, such as AI and machine learning, to effectively manage and manipulate complex systems. This includes creating models and simulations to understand the system's behavior, sensing and acquiring real-time data, preprocessing and analyzing the data, making decisions based on the analyzed data and system models, adapting control strategies in real time, facilitating human-machine interaction, monitoring performance, and optimizing control strategies. The goal is to improve the efficiency, safety, reliability, and overall performance of complex systems in various domains.

AI evolved and developed along with the idea of automatic control, beginning around the 1950s, with the first major applications in computing and information science, and later in automated control (Yev et al., 2000; Russell and Norvig, 1995; Pupkov, 1996). The first commercial and industrial applications of AI can be traced back to the 1980s. AI has attained a certain level of stability and maturity over this period. The significant development in computer technology's capabilities, including hardware implementation of logical and other AI means, is a key aspect that can lead to a reassessment of today's successes and new advancements in AI theory and practice. The phrase "intellectual control system" refers to any combination of hardware and

DOI: 10.1201/9781032659220-4

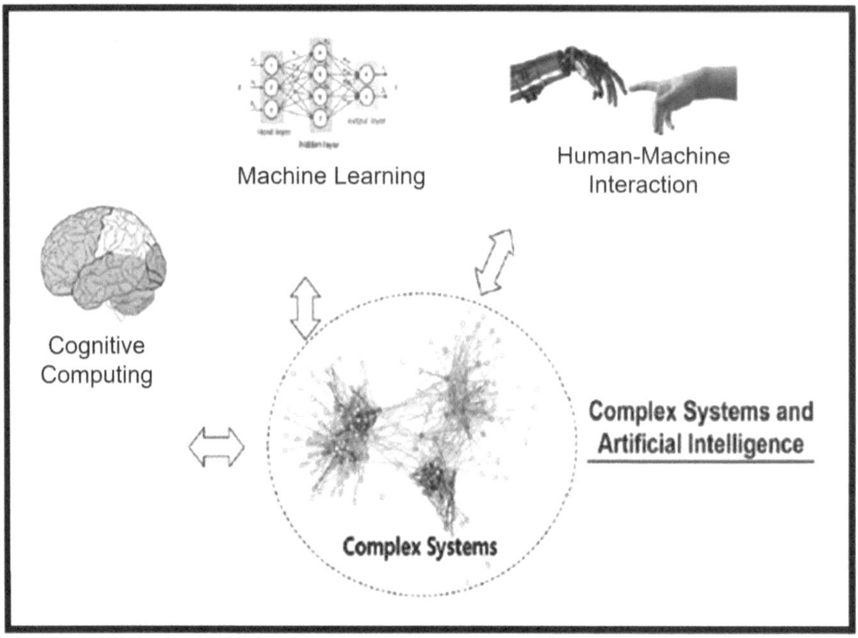

FIGURE 4.1 Example of a Complex System.

(Adapted from Jubair et al., 2024).

software that is linked by a general information process and capable of synthesizing the control objective and finding rational solutions to attain the control goal (in the presence of motivation and knowledge, which includes environmental and internal status information). Human-machine interaction now enables the capacity to synthesize the control goal, and autonomous control systems capable only of finding reasonable solutions to fulfill the control goal are referred to as "intelligent control systems." Currently, the science and practice of control are highly interested in the integration of traditional automatic control methods with AI approaches, as well as AI applications in the field of control for complex, weakly structured objects and processes (see Figure 4.1). Especially when the information, system state, control criteria, and control goals vary over time and become unclear and occasionally contradictory. The report considers a hierarchy of levels of intellectual control as well as a comparative analysis of various AI methods. Because there has been a significant increase in the number of theoretical and applied research in the field of fuzzy controllers over the last decade, the main objective of the study is to examine the key achievements in this area. Unfortunately, even this field does not allow for a thorough review free of the writers' preferences.

Intelligence control for complex systems refers to the ability to effectively manage and manipulate a complex system using intelligent algorithms and techniques. This involves utilizing AI, machine learning, and other computational methods to optimize the functioning of complex systems. There are several key components of intelligence control for complex systems as shown in Figures 4.2 and 4.3.

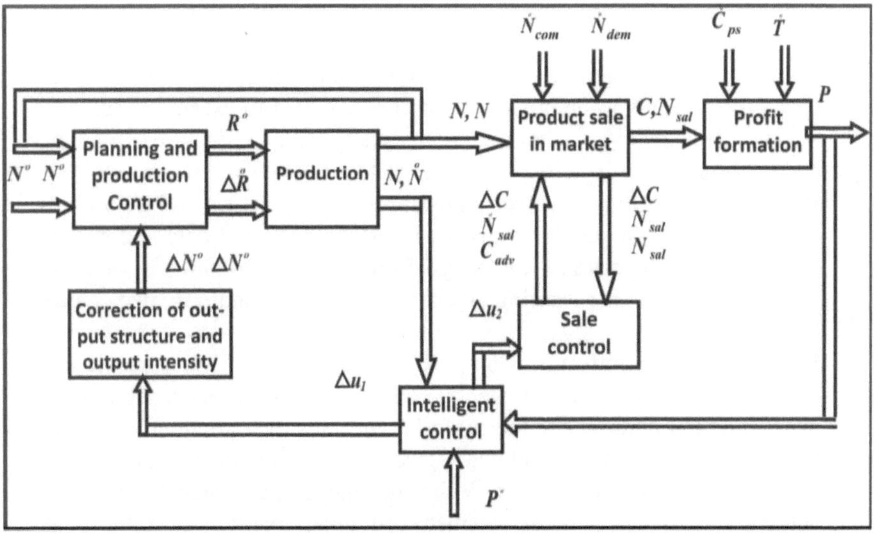

FIGURE 4.2 Intelligent Control.

(Adapted from Jubair et al., 2024).

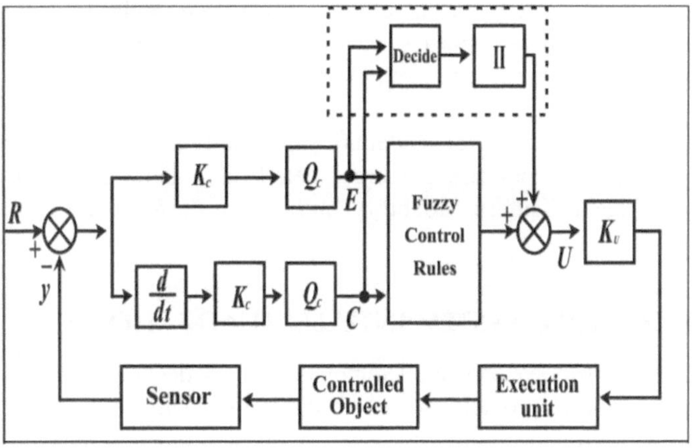

FIGURE 4.3 Fuzzy Control System.

(Adapted from Jubair et al., 2024).

MODELING AND SIMULATION

Creating accurate and detailed models of the complex system is essential to understand its behavior and dynamics. These models can then be used to simulate different scenarios and assess the impact of various control strategies.

- **Sensing and data acquisition**: Gathering real-time data and information from the complex system is crucial for intelligence control. This can involve using sensors, IoT (Internet of Things) devices, and other data collection methods to obtain information about the system's state, performance, and environment.

- **Data preprocessing and analysis**: Raw data collected from the complex system needs to be preprocessed and analyzed to extract meaningful insights. This involves techniques, such as data cleaning, feature extraction, and statistical analysis to transform raw data into relevant information.
- **Decision-making and control algorithms**: Intelligent control algorithms are developed to make decisions based on the analyzed data and system models. These algorithms can range from simple rule-based systems to sophisticated machine learning algorithms that adapt and optimize control strategies over time.
- **Feedback and adaptation**: Complex systems often exhibit dynamic and evolving behavior, requiring adaptive control strategies. Feedback mechanisms are incorporated into the control algorithms to continuously monitor the system's performance and make necessary adjustments in real time.
- **Human-machine interaction**: While the intelligence control algorithms automate many tasks, the involvement of human operators is still critical. Effective human-machine interaction interfaces are developed to facilitate collaboration and decision-making between humans and intelligent control systems.
- **Performance monitoring and optimization**: Monitoring the performance of the complex system and the effectiveness of the intelligence control is essential. Performance metrics and optimization techniques are used to evaluate the control strategies and identify areas for improvement.

Intelligence control for complex systems is applicable in various domains, such as manufacturing, transportation, energy, healthcare, and finance. It can help improve the efficiency, safety, reliability, and overall performance of these complex systems.

GENERAL CONTROL SYSTEM INTELLECTUALIZATION ISSUES

The efficacy of control systems and technologies being developed is critical to the successful resolution of problems to ensure the state's technological independence in the sphere of civil and military purpose, complex technical object creation, and application. There is adequate theory and control technology, taking into consideration the possibility of a lack of specific (depending on the application) required resources: information, timing, energy, money, material, human, and so on. Accidents and disasters in transportation, industry, and energy are frequently related to the so-called "human factor" (HF), including operator overwork. HF often occurs as a result of quality issues in control system design, particularly in controllability emergencies. Human errors, as well as the exhaustion of technical resources of objects and control systems, are all too common in today's Russian environment. They urgently require guaranteed control dependability and quality, along with updates to project, operational, and modernization control capacities. Under the following conditions, methodologies and technologies for evaluating control systems and ensuring their optimality, functional and operational dependability, efficiency, fault tolerance, and survivability are required:

- A lack of prior knowledge about the control object and its external environment of operation, even under opposition conditions,
- A large number of stationarity elements that are difficult to account for due to their subjective nature,
- Degeneration (due to failures or accidents) or the requirement for focused reconfiguration (revitalizing or developmental control).

Control systems become significantly more complex as the functional burden increases. Among the complexity factors of modern and advanced control systems are the following:

- Multilevel controls, heterogeneity in subsystem description by quantitative and qualitative models, different scales of processes in space and time, multimodality, multilink, decentralization, ramified nature, and general structural complexity of modern control systems and their control objects,
- Presence of uncontrolled coordinate-parametrical, structural, regular, and singular impacts, including active counteraction in a conflict environment,
- The use of non-linearity, dispersed parameters, delays in control or object dynamics, impulsive impacts, high model size, and other factors.

Adaptive, robust, predictive, and other control approaches established in control theory are designed to address the incompleteness of dynamics by receiving missing information either during the training stage or in real time. The use of AI involves extending the capabilities of complex control systems by addressing tasks that are unknown or quantifiable. This includes models that became invalid after a certain point in time, and tasks where quantitative models are less efficient than using AI models (such as in action planning tasks) or can be used in conjunction with AI models (Yev et al., 2000). For action planning tasks and control in general, a range of AI approaches such as neuronal, evolutionary, logical, and others can be applied.

Adaptive, robust, predictive, and other control approaches established in control theory are designed to take into account. Each of these classes has its advantages and limitations, particularly in terms of real-time requirements, and ensures the implementation of higher levels of heterogeneous control over complex systems. The intensive development of technical systems and technological processes (networking, miniaturization of sensors, controlling devices and calculators, improving their performance, and so on) places new demands on modern control systems and creates new opportunities. These opportunities arise both at the level of embedded control systems of various scales and at the level of group interaction of decentralized multi-agent systems. The shift from robots operating in an unpredictable environment yet equipped with an operator interface. Current research and development is focused on the transition from robots operating in an unpredictable environment equipped with an operator interface (supervisory Unmanned Aerial Vehicles (UAVs)) to intelligent robots. At that point, less expensive robots based on a modular concept of construction and miniaturization are required to solve sensitizing, environment modeling, robot control team goals, and application scope extension difficulties. Even in agricultural and road construction, drastic standards transformation demands robots with high-precision navigation and cognitive control.

Large-scale infrastructure systems in the electric power industry are examples of critically essential technological processes and intellectual control objects. In this context, several key challenges are evident:

- An inefficient structure of electrical grids and producing capacity,
- A lack of energy savings in electricity usage,
- Technological and commercial losses in electric networks,
- Technological lag and excessive equipment wear,
- A high level of monopolization in power markets, and
- Vulnerability of electric power networks to terrorist and cyber attacks.

Other factors necessitate the development of models of complex infrastructure dynamic systems and the development of efficient solutions and highly dependable intelligent control systems for smart grids (Vassilyev et al., 2011; Berdnikov et al., 2012). Control-based oenological-reactive (production) knowledge model in the so-called expert, recommender, or systems that help decision-making needs to be improved with new features as follows:

- Collaborating with other intellectualization techniques for control systems (artificial neural networks; genetic algorithms; and adaptive, resilient, and predictive control algorithms),
- By combining methods of symbolic and multimedia presentation and knowledge processing,
- Operating with partially formalized and natural language texts, abdicative, and inductive knowledge updating,
- Integrating quantitative and qualitative models with the ontologies of various subject domains that characterize the problem situation, logical control systems' interface complexity with the external physical world can be reduced.

There are numerous methods to combine various AI techniques. For instance, the 1st order of logical techniques of intellectual control (Yev et al., 2000; Vassilyev, 2008) can be linked with the neuro- reactive and logical-reactive (productional) AI means. While the first two strategies promote "reasonable" conduct, the latter methods can address a wider range of information. Verification of knowledge presentation is especially important at the logical-reactive level (sometimes with its many "if-then" rules). The verification of a knowledge base can be reduced to the dynamic analysis of automata networks in the case of production rules of the Boolean type and constructive semantics. This analysis is further condensed in the class of monotonous automata with respect to the state, through the application of the method of mathematical model properties transfer (Fedunov, 2016).

The key problem in AI is the problem of automatic estimation of irrelevance of knowledge, as both a deficit and a surplus of information can lead to the degradation of intellectual control systems. Recent advances in the field of intellectual control include the automation of searching for ways to achieve the control goal given externally, while the automation of goal-setting and revision of control quality criteria remains insufficient. It is now also recognized that improvement of only "machine components" in developed human-machine systems is not enough for the desired essential increase of their use of the efficiency. This goal in creating anthropocentric systems can be achieved by directing the efforts of engineers and scientists on improving the intellectual component of the "system-core" in the anthropocentric system as a built-in set of algorithms for embedded computers together with algorithms of operator activity, referred to as "on-board intelligence" (Fedunov, 2016; Zheltov and Fedunov, 2015). First and foremost, on-board intelligence is required in aviation, especially in combat situations, typical for fighters, i.e. in the circumstances of the most aggressive external environment and tight timing constraints for the crew. On-board intelligence is a functionally integral complex aimed at the fulfillment of all aircraft tasks (Zheltov and Fedunov, 2016). Scientific and technological advances in this field will also be useful for other applications of AI in the conditions of multicriticality, uncertainty, and risk to improve control quality in a situation of information overloading the operator, limited time, or stress. The development of practically useful on-board decision-making support expert systems, including those based on fuzzy logic and case-based reasoning by analogy, has reached the practical stage of building models and prototypes. They are intensively developed in the world to support the creation of the manned combat aircraft of the 4th and 5th generations, as well as combat UAVs. Their fragments already appear on the modernized fighters of 4th generation. In foreign developments, they are planned to be used, first of all, to on board the new USA fighters F-22, F-35, modernized F-16, F-15, F/A-18, and helicopters, which have several on-board intellectual systems of tactical decision-making (Zheltov and Fedunov, 2015). The results of the research, the improvement of on-board computers, cockpit displays and controls, and other avionics give the constructors of next-generation aircraft/helicopter an opportunity to design and realize a-board computer systems of a new type. These systems will be capable to support tactical decisions making (the prompt appointment of the current purpose of flight and choice of a rational way of achieving the goal). Solving such tasks on past-generations aircraft could only be completed by the efforts of the crew. Further, we consider in detail some questions of intellectualization of automatic control systems in the form of fuzzy regulators and combining them with other AI means. Note that the first regulators developed in Greece in the 3rd century BC can partly be considered as the fuzzy controllers described linguistically with logical operations. Today, a huge number of practical applications of fuzzy control systems in the industry, transport, energy, oil and gas, metallurgy, medicine, and other industries and household appliances can be observed in Japan, China, USA, Germany, France, Britain, Russia, and other countries. We consider four basic types of regulators: logical-linguistic, analytical, learned, and proportional-integral differential fuzzy controllers (Kudinov, 1990; Kuzmin and Travkin, 1992; Zakharov and Ulyanov, 1992; Zakharov and Ulyanov, 1993a, 1993b; Zakharov and Ulyanov, 1994; Zadeh, 1976). Since the information about them is unsystematic and is scattered in many publications, our analysis will help a specialist to orient himself in this field.

Intelligence control for complex systems involves employing advanced algorithms and techniques, including AI and machine learning, to effectively manage and manipulate complex systems. The key components of intelligence control include modeling and simulation, data acquisition and preprocessing, decision-making and control algorithms, feedback and adaptation, human-machine interaction, and performance monitoring and optimization. By using intelligence control, we can better understand the behavior of complex systems, make informed decisions based on real-time data and system models, and adapt control strategies in real time. This can lead to improved efficiency, safety, reliability, and overall performance of complex systems in various domains.

Intelligence control is a vital area of research and development as we continue to face increasingly complex systems in our modern world. With advancements in technology and the availability of large amounts of data, intelligent algorithms have the potential to greatly enhance the management and control of these complex systems, leading to better outcomes and more efficient operations.

REFERENCES

Berdnikov RN, Bushuev VV, Vassilyev SN, etc. *The Concept of Intellectual Power System of Russia with Actively Adaptive Network*. V. E. Fortov, A. A. Makarov. M.: JSC FGC UES; 2012. 235 p.

Fedunov BE. On-board operatively advising expert systems of tactical level for manned aircraft. *International Aerospace Magazine "Aviapanorama"* 2016; 1: 9–20.

Jubair MD et al. (2024), "A Comprehensive Review on Intelligence Control for Complex System," *International Journal of Advanced Engineering Research and Science (IJAERS)*, Vol. 11, Issue 1, pp. 44–49, January, 2024, https://ijaers.com/. https://doi.org/10.22161/ijaers.11.7

Kudinov YI. Fuzzy control system, Izv. ussr academy of sciences. *Technical Cybernetics*., Vol. 5, 196–206, 1990.

Kuzmin VB, Travkin SI. Theory of fuzzy sets in control problems and principles of the device of fuzzy processors: A Review of foreign literature. *Automatics and Telemechanics*., Vol. 11, 3–36, 1992.

Pupkov KA. Some results of development of scientific- technical program *"Intellectual Systems" of FKP "Universities of Russia"//Proceedings of the 2 International. Symposium "Intellectual Systems" (INTELS'96)*, Vol. 1, 5–9, 1996.

Russell SJ, Norvig P. *Artificial Intelligence: A Modern Approach*. Prentice-Hall, Inc., 1995.

Vassilyev SN. Formalization of knowledge and control on the basis of positively-formed languages. *Information Technologies and Computing Systems*, Vol. 1, 3–17, 2008.

Vassilyev SN, Yadykin IB, Morgin YuI, Voropai NI, Efimov DN. Intelligent operating and emergency control technologies in Russia CD *Proceedings of the 2th IEEE PES International Conference and Exhibition on «Innovative Smart Grid Technologies – Europe»* (ISGT 2011 Europe, Manchester). Manchester: University of Manchester, 2011. p. 1–7.

Yev SN, Zherlov AK, Fedosov EA, Fedunov BE. *Intelligent Control of Dynamical Systems*. M.: Fizmatlit; 2000. 352 p.

Zadeh LA. *The Concept of Linguistic Variable and Its Application to Making Approximate Decisions*. M.: Mir, 1976. 165 p.

Zakharov VN, Ulyanov SV. Fuzzy models of intelligent industrial controllers and control systems. I. Scientific- organizational, technical, economic and applied aspects, Izv. *Russian Academy of Sciences*. Technical Cybernetics. Vol. 5: 171–196, 1992.

Zakharov VN, Ulyanov SV. Fuzzy models of intelligent industrial controllers and control systems. II. Evolution and principles of construction, Izv. *Russian Academy of Sciences*. Technical Cybernetics.Vol. 4: 189–205, 1993a.

Zakharov VN, Ulyanov SV. Fuzzy models of intelligent industrial controllers and control systems. III. Design methodology, Izv. *Russian Academy of Sciences*. Technical Cybernetics.Vol. 5, 197–220, 1993b.

Zakharov VN, Ulyanov SV. Fuzzy models of intelligent industrial controllers and control systems. IV. Simulation modeling, Izv. *Russian Academy of Sciences*. Technical Cybernetics., Vol. 5, 169–211, 1994.

Zheltov SYu, Fedunov BE. Operational goal-setting for anthropocentric objects from the point of view of the conceptual model "Etap", I, II Izv. *Russian Academy of Sciences. TiSU*. Vol. 3, 57–71, 2015; 2016; 3, 55–69.

5 Modeling of Kanban Production Systems

BACKGROUND

Based on Srivathsan and Kamath (2014), a production system could be viewed as a network of manual or automated workstations that can perform various operations to convert raw material into a finished product. A workstation could also include an output store where a limited amount of completed parts is stored. After the completion of processing at a workstation, a part can be sent to the next workstation in the part's routing or temporarily stored in the output store of the workstation and released based on a signal from a downstream workstation. The former is a typical characteristic of a push system, and the latter of a pull system. Production systems in which the production start date for a product is determined by subtracting the production lead time from the due date that has been forecasted by a material requirement planning (MRP) tool tend to use the push material flow control strategy. Production systems that are driven more by actual customer demands tend to operate under the pull material flow control strategy. The often-mentioned advantage of a pull system over a push system is reduced average work-in-process (WIP), both planned and unplanned, resulting in lower inventory holding costs and shorter average lead times.

Various mechanisms have been employed to implement a pull-based material control system such as the Kanban, base-stock, CONWIP, POLCA, extended Kanban, and generalized Kanban. Kanban (Japanese word meaning signboard) is a scheduling system for lean manufacturing (also called just-in-time manufacturing, abbreviated JIT). Taiichi Ohno, an industrial engineer at Toyota, developed Kanban to improve manufacturing efficiency. The system takes its name from the cards that track production within a factory. The choice of a control mechanism by a firm depends primarily on its preference with regard to the trade-off between the cost of holding inventory, cost of backorders, and lost sales. In addition to the push and pull strategies for material flow control, production systems also use different strategies such as make-to-stock, make-to-order, and assemble-to-order for various products that determine what type of and how much inventory is maintained in the system. In make-to-stock, finished or semi-finished products are kept in stock before a customer order is realized, whereas, when the product is make-to-order, no finished goods inventory is stored and products are produced only after a customer order is realized. In an assemble-to-order strategy, the finished product is assembled after customer demand is realized, while the key parts and sub-assemblies used in the assembly process are stocked in anticipation of demand. This strategy is a combination of the make-to-stock and make-to-order strategies.

In today's highly competitive and global marketplace, companies need to ensure that high-quality products are readily available to quickly satisfy customer demand,

 DOI: 10.1201/9781032659220-5

all at a low cost. In such a competitive market scenario, the combination of a pull material flow control strategy to reduce average WIP and manufacturing lead times and make-to-stock strategy to increase the chance of instantaneous demand satisfaction is very relevant. A make-to-stock, pull-controlled production system is referred to as a production-inventory system in the literature (Duri et al., 2000). The focus of this chapter is on the performance modeling of production and inventory systems. It provides (i) a brief overview of the various pull-based material flow control mechanisms, (ii) a review of quantitative models available to analyze the performance of production-inventory systems under various pull-control mechanisms, and (iii) a comprehensive list of references on modeling production-inventory systems for academic researchers and industry practitioners.

PERFORMANCE EVALUATION MODELS

Performance evaluation models serve as an aid to system designers and operations managers. These models are useful in obtaining some of the key performance measures of a system for a given set of parameter values (e.g. type and number of resources, demand and processing rates, variability in processing times, etc.) through the development and solution of analytical/simulation models (Suri et al., 1993). Additionally, these models aid in understanding the complex dynamics between the various system parameters and in taking decisions on key business parameters for the system such as the target inventory for raw materials, intermediate goods, and finished goods as well as capacities of production resources. The common tools that can be used in performance evaluation include Markov chains, stochastic Petri nets, queueing networks, and simulation models (Leung and Suri, 1990; Srivathsan, 2012). Leung and Suri (1990) provided a brief overview of these performance evaluation techniques and also discussed the situations in which each of these tools could be employed. These performance evaluation tools can also be coupled with optimization models or used in a "what-if" mode to support key tactical design decisions (e.g. Liu et al., 2004).

KANBAN MECHANISM

The Kanban control mechanism is one of the best known pull-control mechanisms that dates back to the mid-seventies when it was originally developed by the Vice-President of Toyota Motor Company, Mr. Taiichi Ohno, and employed in the Toyota Production System (Sugimori et al., 1977; Ohno, 1988). Kanban cards are used to control the release of parts at each stage of production. In the cases of multiple product types, separate kanban cards are assigned for each product type at each production stage. The number of kanban cards (for each product type) at each stage would limit the WIP (for that product) at that stage (Duri et al., 2000). Each individual stage of a kanban-controlled mechanism has a fixed number of kanban cards. A part is released from the output store of stage $i+1$, P_{i+1}, to stage i only when a kanban card authorizing production order at stage i is available at D_{i+1}. When the kanban card is available at stage i+1, D_{i+1}, and the product is available at stage i+1, P_{i+1}, the stage i kanban card is attached to the product at stage $i+1$, P_{i+1}, and released to the workstation at stage *i* for processing. The processed product from stage *i*, P_i, would be released to the workstation at stage *i*-1 when a kanban card authorizing production at stage *i*-1 becomes available at stage *i*, D_i. At this time, the stage *i* kanban card is removed from the product and sent to stage *i*+1, D_{i+1}, as production authorization for stage *i*. Thus, the kanban cards provide a means to combine both demand information and material flow in the production system.

Kanban systems are classified as single-card Kanban systems and dual-card Kanban systems. The single-card Kanban system is similar to the Kanban system described above in which the production card is used. In a dual-card Kanban system, two types of cards namely the withdrawal cards and the production cards are used. The production cards work in a way similar to those in a single-card Kanban system, while the withdrawal cards are used for material requisition. In this system, when demand consumes an item from an output store at stage *i*, the withdrawal card is removed and sent to the output store at the preceding stage (*i*+1). When the withdrawal kanban from stage *i* and a product is available at stage *i*+1, a part is released to the workstation at stage *i* for processing. The production kanban attached to this released part is then sent as a production authorization card to stage *i*+1, and this results in the replenishment of inventory at the output store of stage *i*+1. For a detailed overview of Kanban systems, the reader is referred to Huang and Kusiak (1996).

Analytical Models of Kanban Control Systems

Since the 1980s, there has been considerable interest in studying the behavior of kanban systems. Various performance evaluation tools like simulation, Petri nets, Markov chains, and queueing models as well as optimization tools have been employed for this purpose. This chapter presents a brief review of some of the performance evaluation models developed for Kanban systems.

The early studies in the field focused on the development of simulation models to study the effect of the production environment on the performance of the Kanban system as well as to compare the performance of push and pull-controlled production systems. Huang et al. (1983) developed simulation models of a Kanban system for a multi-line, multi-stage production system with an interest in determining if the system can be employed in the American industry where the production environment was characterized by variable processing times and imbalances between the production stages. The study led to the conclusion that the Kanban system could not be employed in the US without making changes to the production environment such as standardizing the processing and setup times through "substantial amount of worker training and cross-training (Huang et al., 1983)." Krajewski et al. (1987) conducted a simulation study to ascertain the areas of the manufacturing environment in the US that needed the most attention as well as to assess the robustness of the Kanban system to various production environments in the US. The study revealed that the major factors that needed to be addressed included lot sizes, setup times, yield losses, workforce flexibility, degree of product customization, and product structure, and that the performance of the Kanban system was dependent on the production environment. Sarker and Fitzsimmons (1989) used a simulation model to compare the performance of push and pull systems under the effect of variability of operator performance and unequal distribution of task times. As per their study, the pull system performed better than the push system with respect to minimizing the WIP. It was also seen that the pull system was more sensitive to higher variability in processing time, leading to a conclusion that "lower variability in operation times and the even allocation of workloads along the line is advisable to achieve a better output rate in a pull system (Sarker and Fitzsimmons, 1989)." Di Mascolo et al. (1989) presented a basic Petri net model that provided a unified approach to modeling Kanban systems and analyzed a single-stage Kanban system using an approximate method. The study found that the production capacity of the system increased with an increase in the number of kanban cards and decreased with an increase in variability in processing times.

Deleersnyder et al. (1989) employed a discrete-time Markov chain model to analyze the effect of various system parameters on the performance of a general N-stage serial Kanban-controlled production system. They defined the system state at the end of each period by N parameters representing the inventory level at each stage. The demand was modeled as a stationary binomial distribution. The study revealed that lower variability in demand and high machine reliability maximize the benefits of the system, although higher demand variabilities and frequent machine breakdowns could be offset by increasing the number of kanban cards, thus increasing the inventory. The study also revealed that adding safety stock was more appropriate when the safety stock was less than the minimum backlog as obtained from the steady-state probability mass function of the backlog in the system. Similarly, safety stock greater than the maximum backlog was undesirable, while trade-offs needed to be made between the inventory level and a system service level (e.g. fill rate, the relative importance of the inventory and backlog costs, etc.) when the safety stock was between the maximum and minimum levels. Karmarkar and Kekre (1989) used continuous-time Markov chain (CTMC) models to study single-stage kanban cells with single and dual cards as well as two-stage Kanban systems. The CTMC model was numerically solved to study the effect of batch sizing on the system costs comprising of the inventory holding and the backorder costs. Their study showed that the performance of the system was significantly affected by both the number of cards and the batch size. It was also noticed that the effect of the number of cards on the costs was more symmetrical than the effect of the batch size on the costs (Karmarkar and Kekre, 1989). Mitra and Mitrani (1990) studied an N-stage tandem Kanban system, where each stage had a single processor with exponential processing times. They assumed an ample supply of raw materials at the first stage and a similar pool of demand at stage N. They used an approximation scheme to solve this system by decomposing the multi-stage system into subsystems consisting of the individual stages. They examined the individual subsystem in isolation using a CTMC model where the state was defined as the vector of integers representing the number of cards and the number of parts. Then, an iterative procedure was used to analyze the subsequent subsystems. The results of the study showed that the Kanban system is equivalent to a tandem network consisting of finite-buffer, single-server workstations operating under the minimal blocking policy. Mitra and Mitrani (1991) extended this model to incorporate stochastic demand.

Di Mascolo et al. (1996) considered an N-stage serial Kanban-controlled production system with a single product type and infinite raw material supply. The N-stage system was decomposed into N subsystems, each representing a single-stage Kanban system, consisting of the processing node in the stage and two synchronization stations (one upstream and one downstream). Each subsystem was synchronized with two external arrival streams (assumed to follow a state-dependent Markovian process) representing demand from the downstream stage and raw material supply from the upstream stage. Each subsystem, modeled as a CQN (Channel Quality Node) with two synchronization stations, was approximated into a product form network by replacing the processing nodes at the stations with exponential stations with load-dependent processing times. The accuracy of the approach was tested exhaustively by varying the number of stages, the number of cards at each stage, the topology, and the processing time distribution. The analysis showed that the production capacity of the saturated Kanban system was accurately estimated except when the number of kanban cards at the various stages was small. Similarly, the errors in estimating measures like mean waiting time for a back-ordered demand and stock-out rate of the Kanban system with back-ordered demand were small except when (1) the number of kanban cards was small, (2) the demand rate was close to the production capacity, or (3) the processing time SCV (Strategic Control Variable) was high. Such insight into the performance prediction capability of the models developed could be useful when the models are used as aids in designing/operating real-world systems.

Krishnamurthy (2002) developed queueing models for the performance evaluation of a wide set of material control strategies in production systems based on the parametric decomposition approach and two-moment queueing approximations. Krishnamurthy and Suri (2006) employed the parametric decomposition approach for the performance evaluation of Kanban-controlled production systems with general demand processes, material arrival processes, and processing times. A multi-stage Kanban system was decomposed into single-stage Kanban systems that served as a building block. The building block was modeled as a CQN with a processing stage and two fork/join (or) synchronization stations. The arrival process and processing times at each stage were assumed to follow a renewal process. Each stage was studied in isolation to obtain the departure process from the stages as well as the performance measures for the stages. A set of non-linear equations was obtained by linking the traffic processes at the individual stages, and these equations were solved using an iterative procedure to determine the unknown parameters characterizing the internal traffic processes. This information was then used in the calculation of the network performance measures, such as system throughput and mean queue lengths at each stage. The study revealed that system throughput was nearly insensitive to change in variability at one or more stages or processes, while system-wide variability resulted in much lower system throughput.

Krishnamurthy and Ge (2009) investigated the benefits of integrating advanced demand information (ADI) with the Kanban control system using a Markov chain model. They assumed that the customers placed their orders in advance of the due date by a period called the demand lead time (assumed to be exponential). The state of the Markov chain was defined as $(F(t) - P(t), I(t) - B(t))$, $t \geq 0$, where $F(t)$ is the number of free kanban cards available, P(t) is the number of pending orders waiting for free kanban cards, I(t) is the number of finished goods, and B(t) is the number of backorders. The Chapman-Kolmogorov equations were then solved to obtain performance measures such as the stock-out rate, expected inventory level, expected number of backorders, and average WIP. The Kanban system with ADI was compared with a pure Kanban system and the base-stock policy with ADI, and the results revealed that the Kanban system with ADI resulted in lower inventory and backorder levels.

BASE-STOCK MECHANISM

The study by Clark and Scarf (1960) showed that the base-stock policy is an optimal policy for an uncapacitated manufacturing system with constant lead times and serves as a classical reference for research on base-stock policy. The base-stock policy is defined by a parameter at each stage called the base-stock level, which represents the maximum planned inventory at that stage. Unlike a Kanban system, the external customer order at the last stage triggers an instantaneous replenishment order at every upstream stage in the network. As a result, the maximum WIP in a base-stock controlled production system is not bounded (Duri et al., 2000). This may lead to localized WIP build-up at a workstation. However, the WIP queue length at each stage is bounded by the base-stock level. The advantage of such a control mechanism over the kanban policy is that there is no blocking of demand information, thus providing the workstations more time to prepare and react. This can lead to a higher service level when compared to the Kanban system. It also has to be noted that the base-stock policy is identical to the kanban policy for a single-stage production system. An arrival of customer demand at stage 1, D_1, triggers an instantaneous replenishment order at each of the downstream stages 2, 3, and 4 (D_2, D_3, D_4). If a finished product is available at the output store of stage 1, P_1, then the arriving customer order is immediately satisfied. Similarly, when a semi-finished product is available at the output store of stage $i+1$, P_{i+1}, it is immediately released to stage i to satisfy a replenishment order placed by stage i at stage $i+1$, D_{i+1}.

ANALYTICAL MODELS FOR BASE-STOCK SYSTEMS

Gavish and Graves (1980, 1981) studied a single-stage, single-product base-stock system with Poisson demand. Gavish and Graves (1980) used an M/D/1 queueing system to find the control policy that minimized the expected cost per unit time, while Gavish and Graves (1981) used an M/G/1 queueing system to accomplish the same. The optimal decision policy for both the models was found to be a two-critical-number policy characterized by the parameters (Q^*, Q^{**}), "such that the server is turned off when the queue length is first reduced to Q^*, and is turned on when the queue length first reaches Q^{**} (Gavish and Graves, 1980)." Graves and Keilson (1981) studied a single-product, single-machine problem under Poisson demand and exponentially distributed order size. They minimized the system cost that comprised setup cost, inventory holding cost, and backorder cost using a spatially constrained Markov process model.

Buzacott and Shanthikumar (1993) presented results for general single-stage systems with unit demand and backlogging using the production authorization (PA) card concept, where a tag attached to an item is converted into a PA card when demand consumes the item. Using the number of outstanding orders in the stage at time t, K(t), and the total number of tags in the system, *S*, they obtained the inventory level at time t, I(t), and the number of customer orders back-ordered at time t, B(t) using equations (5.1) and (5.2). Since the PA card represents an outstanding order at a stage, the number of PA cards available at time t, C(t), can be obtained using equation (5.3). It can be seen from equations (5.1) through (5.3) that by studying the process K(t), we can derive I(t), B(t), and C(t). The fundamental relationship between the expected number of outstanding orders, the total number of tags, the expected inventory level, and the expected number of orders back-ordered is given by equation (5.4) (Svoronos and Zipkin, 1991).

$$\mathrm{I}(\mathrm{t}) = \max\left[0, \mathrm{S} - \mathrm{K}(\mathrm{t})\right] \tag{5.1}$$

$$\mathrm{B}(\mathrm{t}) = \max\left[0, \mathrm{K}(\mathrm{t}) - \mathrm{S}\right] \tag{5.2}$$

$$\mathrm{C}(\mathrm{t}) = \min\left[\mathrm{S}, \mathrm{K}(\mathrm{t})\right] \tag{5.3}$$

$$E\left[B\right] = E\left[I\right] + E\left[K\right] - S \tag{5.4}$$

Buzacott and Shanthikumar (1993) also modeled single-stage systems with lost sales, interrupted demand, bulk demand, machine failures, and yield losses. These models have been used as a building block by Sivaramakrishnan (1998) and Srivathsan (2004).

Benjaafar et al. (2004) examined the effect of product variety on the inventory cost in a finite capacity production-inventory network where the products were manufactured in bulk and shared a common manufacturing facility. The production facility was assumed to comprise a batching stage and a processing stage. The production stage was viewed as a GI/G/1 queue whose arrival process was the superposition of the order arrivals for the individual products. The key findings of the study were that the total cost increased linearly with the number of products and that the rate of increase depended on the system parameters such as demand and processing time variability, and capacity levels.

Multi-stage Systems

The analytical models for multi-stage tandem base-stock systems developed by Lee and Zipkin (1992), Sivaramakrishnan (1998), Liu et al. (2004), and Srivathsan (2004) have common assumptions, such as one-for-one replenishment with back-ordering permitted at each stage, no limit on WIP queues at a stage, and ample raw material supply at the first stage (Kamath and Srivathsan, 2012). The last three approaches are based on the parametric decomposition approach (Whitt, 1983, 1994) and share a common solution structure as explained in Kamath and Srivathsan (2012). The performance measures that were of interest in the four approaches are the fill rate (the probability that an order is satisfied instantaneously), the expected inventory level, and the expected number of backorders.

Sivaramakrishnan and Kamath (1996) modeled an M-stage tandem base-stock system by decomposing it into individual stages where each stage is made up of the manufacturing resource at that stage and a delay node to "capture the upstream delay experienced by an order when there was no part in the output store of the previous stage (Sivaramakrishnan, 1998)." Sivaramakrishnan (1998) modeled the manufacturing unit as a GI/G/1 queue and the delay node as an M/G/∞ queue. Sivaramakrishnan (1998) extended the tandem model to include (i) multiple servers at a stage, (ii) batch service, (iii) limited supply of raw material, (iv) multiple part types, and (v) service interruptions due to machine failure. Sivaramakrishnan (1998) also extended his approach to tandem networks with feedback and feed-forward networks. The main contribution of this study was the use of the delay node as a building block construct. Liu et al. (2004) modeled a multi-stage tandem manufacturing/supply network with general inter-arrival and service processes. The tandem network was modeled as a system of inventory queues and then the overall inventory in the network was optimized with service-level constraints. A decomposition scheme was proposed, in which a semi-finished product from a stage is moved into the input buffer of the downstream stage, whenever the job queue at a stage (material queue at the input buffer of the stage + backorder queue at the stage) is increased by one. If the inventory store at the upstream stage is empty, then the request for the semi-finished part is back-ordered. As a result, the job queue or outstanding orders queue at a stage consists of the material queue comprising of the semi-finished products from the upstream stage and a backorder queue. In this approach, the backorder distribution is used to account for any delay due to stock-out at an upstream stage. The approximation was found to yield better results at higher base-stock levels and at higher fill rates.

Svoronos and Zipkin (1991) modeled a multi-echelon inventory system with Poisson demand arrivals at the lowest hierarchy of the system called the leaf. Each location was assumed to operate under the base-stock policy, resulting in instantaneous propagation of demand from the leaf to the central depot. The central depot was assumed to have infinite raw material supply. Each location was considered to be a node, and each arc connecting a node with its predecessor was treated as a transit system (representing transportation or production activities). They assumed that parts are processed sequentially and the behavior of each transit system was assumed to be independent of the demands and orders in the inventory system. The transit systems for all the arcs on the path from the outside source node to the leaf node were assumed to be mutually independent. Svoronos and Zipkin (1991) applied the results of the single-location problem recursively, starting from the highest-level echelon to analyze the complete network. Lee and Zipkin (1992) modeled a tandem production-inventory network with Poisson demand arrivals and exponential service times at all stages. The multi-echelon system developed by Svoronos and Zipkin (1991) was used to analyze the tandem system by assuming that stage i behaved like an M/M/1 queue (Lee and Zipkin, 1992), and the sojourn times were exponentially distributed with mean $1/(\mu_i - \lambda)$, where λ is the demand rate, and μ_i is the processing rate at stage i. The sojourn times at all the stages were considered to be independent. Such an effective lead-time decomposition approach is similar to Whitt's parametric decomposition approach (Whitt, 1983). The approach developed by Svoronos and Zipkin (1991) that employed phase-type approximations (Neuts, 1981) was then used to solve the above approximate system. The approximation was found to perform well over a wide range of parameters like inter-arrival and processing times and base-stock levels. Zipkin (1995) extended this work to model tandem production-inventory networks with feedback.

Kamath and Srivathsan (2012) have identified and documented the similarities and differences between the approaches developed by Lee and Zipkin (1992), Sivaramakrishnan (1998), Liu et al. (2004), and Srivathsan (2004). They also conducted an extensive numerical study to better understand the performance prediction capability of these approximations. The study concluded that the models developed by Liu et al. (2004) and Srivathsan (2004) perform better than the model developed by Sivaramakrishnan (1998) for the system with general inter-arrival and processing time distributions. In the case of the system with Poisson arrivals and exponential processing times where the final stage was make-to-order (experiments conducted by Lee and Zipkin (1992) were replicated), it was concluded that Sivaramakrishnan's model worked best.

Benjaafar et al. (2005) studied the effect of inventory pooling in base-stock controlled production systems with multiple locations having identical costs. The production facility was assumed to have finite capacity, and supply lead times were assumed to be endogenous. They studied the sensitivity of the cost benefits from inventory pooling to system parameters such as service levels, demand and service time variability, and structure of the production-inventory network. For the system with multiple production facilities, they considered three different pooling scenarios. In the inventory-pooling scenario, each production facility was modeled as an independent M/M/1 queue. In the scenario where both inventory and capacity were pooled, the production system was modeled as an M/M/n queue. In the capacity-pooling scenario, the production facilities were modeled as an M/M/n multi-class FIFO (First In, First out) production system. The results of the study showed that the benefit of inventory pooling decreased with utilization, while the benefit of capacity pooling increased with utilization.

Systems with Advanced Demand Information

Karaesmen et al. (2002) assumed that the inter-arrival times and the processing times are geometrically distributed and modeled the system with ADI for contract suppliers. They analyzed the base-stock policy (S, L) with a focus on optimization and performance evaluation of the Geo/Geo/1 make-to-stock queue, where L is the release lead-time, which was used to "regulate the timing of material release into the manufacturing stage (Karaesmen et al., 2002)." Karaesmen et al. (2004) studied the factors that impacted the value of ADI in a single-stage make-to-stock production system operating under the base-stock policy. The production system was modeled as an M/M/1 queue, and the system performance and value of ADI were obtained exactly for customers accepting early delivery as well as customers not accepting early delivery. The study revealed that the value of ADI decreases at extremely high system loads. Further, ADI coupled with early delivery was found to reduce the producer's costs significantly. Karaesmen et al. (2003) advanced the above model by developing a new approximation scheme for an M/G/1 make-to-stock queue with ADI. The study showed that the value of ADI decreased at extremely high system loads and increased with reduction in average as well as variability in the production lead times.

Systems with Periodic Demand

Bhat and Krishnamurthy (2012) considered a single-stage manufacturing system with periodic demand that follows a non-stationary Poisson process. The processing times were assumed to be exponentially distributed. The production system was assumed to have the flexibility to increase the production rate up to a threshold level at a linear cost. Alternatively, additional capacity could be employed at a higher cost. The stationary demand was modeled as a Markov-modulated Poisson process (MMPP), and Markov decision process model was used to study this system. They considered the infinite horizon discounted expected cost and the long-run average expected cost and developed optimality conditions for both. The existence of optimal solutions and optimal policy was also established. The study revealed that the optimal production rate was non-decreasing with the queue length as well as the phase of the MMPP when the Markov-modulated phase transition is stochastically monotone (Bhat and Krishnamurthy 2012). The switching levels based on the queue length where the producer would initiate production or capacity were identified and the expressions to determine optimal production rates at different queue length ranges were provided.

Other Related Systems

Srivathsan (2004) extended the performance evaluation models for base-stock controlled production-inventory networks developed by Sivaramakrishnan (1998) to model a generic supply chain network (SCN) where each manufacturer was assumed to have an input store for raw material storage and an output store for finished product storage. An example network with three suppliers, two manufacturers, and three retailers was used to illustrate the performance prediction capability of the approximations. Srivathsan (2004) used the backorder distribution to establish the link between successive nodes in the network. Each manufacturer was modeled as a single server queue, and the lead-time delays at the raw material suppliers and the transportation delays in the network were approximated by an M/G/∞ model. The study showed that the accuracy of the model suffered at lower base-stock levels. It was also seen that the performance measures (fill rate, expected inventory level, and expected number of backorders) at the various stores are more sensitive to processing time variability than inter-arrival time variability (Srivathsan and Kamath, 2012).

Srivathsan (2012) developed analytical models to study the effect of upstream inventory information sharing in two-echelon SCNs. Two SCN configurations were studied – one retail store and two production facilities and two retail stores and two production facilities. In both these cases, the production facilities were modeled as an M/M/2 queueing system, and the backorder distribution from the M/M/2 queueing system was used to link the inventory levels at the production facilities and the retail store. The study revealed that the value of information sharing is significant at lower base-stock levels and higher utilizations at the production facilities. The value of information sharing was also found to be more sensitive to processing time variability than to inter-arrival time variability.

Jain and Raghavan (2005) considered a production-inventory system with a manufacturing plant and a warehouse that operated under the base-stock policy. The manufacturing system was modeled as a single-stage discrete-time queueing system, while the warehouse inventory was modeled as the input control mechanism for the manufacturing plant. The customer order arrival process was assumed to be Poisson, and the manufacturer was assumed to process orders from the warehouse at fixed discrete time slots. The study focused on determining the optimal base-stock level at the warehouse that minimized the long-term total expected cost per unit of time (one with non-negativity constraints only and a second with service-level constraint). The numerical study showed that as the ratio of the backorder cost to the inventory holding cost is increased, the total cost and the base-stock level also increased.

Arda and Hennet [3] considered an enterprise network where the end-product manufacturer had several potential suppliers for components. Demand at the manufacturer was assumed to follow a Poisson process and was of unit size, while each supplier was assumed to follow an exponential service time and FIFO discipline. The manufacturer was assumed to follow a base-stock policy. The manufacturer placed an order with supplier *i* with a probability α_i as per a Bernoulli splitting process, and this resulted in Poisson arrivals at the suppliers. So, each supplier was modeled as an M/M/1 queue. An optimization model with an objective of minimizing the average cost of the manufacturer (expressed as the sum of mean inventory holding cost and mean back-ordering cost) was used to obtain the Bernoulli probabilities and the optimal base-stock value. They showed that the problem at hand was hard as the convexity of the objective function was not guaranteed and focused on solving a relaxed problem. The study revealed that dispatching orders to multiple suppliers is more beneficial than placing an order with one supplier.

CONSTANT WORK-IN-PROCESS (CONWIP) MECHANISM

The CONWIP mechanism was initially proposed by Spearman et al. (1989, 1990). This mechanism maintains an upper bound on the WIP in the production system, and when the level is reached, new jobs are not allowed to enter the system until a job leaves. It has to be noted that while the Kanban mechanism maintains an upper bound on the WIP at each stage, the CONWIP mechanism maintains an upper bound on the WIP for the entire production system. This means that when the production system consists of a single processor, the CONWIP and Kanban mechanisms are identical. In this mechanism, the finished product inventory is used to satisfy customer demand, while there is no inventory of semi-finished products at the intermediate stages.

An external customer order requests the release of a finished product from the last stage. If a finished product is available, it is immediately released; otherwise, the order is back-ordered. All of the remaining stages essentially behave like a push system with no blocking. This policy depends only on the number of CONWIP cards, and these cards affect the transfer of finished products as well as the demand. The demand is transferred from the last stage to the first stage, and not to any intermediate stages.

ANALYTICAL MODELS FOR CONWIP SYSTEMS

Hopp and Spearman (1991) studied a tandem production line operating under the CONWIP mechanism using a CQN. They assumed that the machines had deterministic processing times, exponential failures, and repairs. They developed an approximate regenerative model (ARM) to approximate the throughput of the system as a function of the WIP level. The performance of ARM was compared to the mean value analysis, and the study showed that ARM was more robust. Duenyas and Hopp (1990) considered a CQN with m exponential servers, n jobs, and no blocking. The study focused on obtaining the standard deviation of cumulative output from a closed queueing network in a specified time interval using an approximation scheme that worked well for a range of situations. It was seen that the approximation was less accurate when the number of machines was high and also at higher bottleneck utilization values. Duenyas and Hopp (1993) studied a CONWIP system with several fabrication stations feeding an assembly station, where both the fabrication station and the assembly station consisted of multiple exponential processors. The system was modeled as an assembly-like queue and approximations were developed for the throughput and average queue length. The study revealed that an increase in machine speed would increase the system throughput. It was also noticed that a bottleneck at the assembly line limited the throughput more than bottleneck at the fabrication line (Duenyas and Hopp, 1993).

Satyam and Krishnamurthy (2008) analyzed a multi-product production system operating under the CONWIP mechanism. The system was modeled as a CQN with synchronization stations and a new approximation scheme based on the parametric decomposition approach extending the work of Krishnamurthy and Suri (2006) to model multi-class networks was developed. The m-stage network was decomposed into m-production stages and $2R$ synchronization stations, where R represents the number of product classes. The arrival and processing time distributions at the stages were assumed to be renewal processes, and each station was analyzed in isolation to obtain the departure process from the stage as well as the performance measures of the stage. These steps are similar to those used by Krishnamurthy and Suri (2006) for a single-class system. The difference arises in the linkage stage where the arrival process at a stage is the superposition of the departure process for the individual products from the previous station. Another important constraint is that for a product class, the sum of the mean queue lengths at the different production stations and the synchronization stations in its route must be equal to its population (the maximum number allowed in the system). As in Krishnamurthy and Suri (2006), an iterative algorithm was then used to solve the set of non-linear equations that were derived from the traffic equations and variability equations that link the mean and SCV of the arrival and departure processes at the various stations. The study revealed that an increase in the variabilities (of the raw material inter-arrival times, demand inter-arrival times, and processing times) resulted in a decrease in throughput and an increase in the levels of WIP inventory and average number of backorders for each product. It was also seen that an increase in the demand arrival rates (when the demand arrival rates < maximum throughput capacity of the system) resulted in an increase in the levels of throughput, WIP inventory, and the average number of backorders for each product.

POLCA MECHANISM

POLCA stands for "Paired-cell Overlapping Loops of Cards with Authorization" and was introduced by Rajan Suri in 1998 (Suri 1998). POLCA is a hybrid push-pull mechanism that combines "the best features of push/MRP systems and pull/Kanban control" (Krishnamurthy and Suri, 2009). In a system controlled by the POLCA mechanism, the release authorization cards that are generated using a high-level MRP system and production control cards called POLCA cards are used to control the flow of orders through different cells. The high-level MRP system generates a release authorization time at each cell after every demand realization. The cell does not start production immediately at the release authorization time but waits for the corresponding POLCA card to become available (Krishnamurthy and Suri, 2009). According to Krishnamurthy and Suri (2009), POLCA systems are more suited for high variability environments.

The POLCA cards control the material movement between the cells and are assigned to pairs of cells. Suppose the routing for a part is from cell 1–3 and 3–6, and then the POLCA cards are assigned to pairs 1–3 and 3–6. The POLCA card is always attached to the part until the processing at the second cell is completed, i.e. the 1–3 card would be attached to the part as it enters cell 1 and continues to be attached till the part moves to cell 3 and finishes getting processed at cell 3. At this point, the card is returned to cell 1, signifying that capacity is available at the downstream cell 1. Similarly, the 3–6 card would be attached to the part when it enters cell 3 and would continue to be attached to the part till the processing at cell 6 is completed. This shows that the POLCA card serves as a capacity signal rather than an inventory signal as is the case with a kanban card.

Vandaele et al. (2008) proposed a framework in which they have linked a load-based POLCA system with an advanced resource planning system (ARP) to study the stochastic behavior of the production system with limited capacity. In this framework, the ARP system was driven by a queueing network model in which each workstation was modeled using a GI/G/1 queue, and the aggregate arrival process and the aggregate service process were obtained as a function of lot sizes. A closed-form expression was obtained for the weighted average lead time for the customers as a function of the lot sizes. An optimization model was then used to minimize the weighted average lead time for customers to obtain the optimal lot sizes for the different products. The optimal lot size and the optimal average lead time were then used in the calculation of the distribution of the production order lead times and the waiting times in queue. These performance measures were then used in the calculation of the authorization times and the allowable workload per loop for the load-based POLCA system. The study led the authors to conclude that the load-based POLCA system offers a promising approach for industrial practice. Vandaele et al. (2008) also present the details of the implementation of such a system at Spicer Off-Highway Products Division in Belgium.

Riezebos (2010) presents an overview of the design issues in POLCA systems and specifically focuses on tools that can be used in the implementation of POLCA systems in make-to-order companies. Riezebos (2010) proposed a five-step scanning procedure that could be used to investigate the suitability of a POLCA system for a firm. In the first step, the original source of the firm's problem is identified and the use of POLCA in mitigating the problem is analyzed. This is followed by a gap analysis to set the objectives for the POLCA. In the second step, the resource and layout characteristics are analyzed to decide on the appropriate POLCA cells, keeping in mind any bottleneck stations and other considerations. In the third step, the key process flow and capacity characteristics like utilization levels, cell throughput time estimates, and the work content of the job are observed to check if there is a need for additional improvements in data accuracy and capacity control. In the fourth step, the product types that need to be controlled using POLCA are identified and a from-to routing chart for the product types through the cells is developed. This information is used to set the actual loop. In the final step, the product mix and the available production flexibility are considered to determine the number of POLCA cards that need to be circulated in each loop. The effectiveness of the POLCA system is then assessed to ascertain if additional improvements need to be made and if the initial objectives have been achieved.

Other variations of the POLCA system like the GPOLCA (`generic paired-cell overlapping loops`) or the generic POLCA have been proposed. The GPOLCA mechanism implements a push control mechanism, and the order release is controlled by a combination of release dates and cards that are allocated to pairs of cells (as in the case of the POLCA mechanism). However, this mechanism does not allow for replenishment of inventory and maintains an inventory of GPOLCA cards to limit the load of the system. An order is released only when the required cards from each pair of cells (in the job routing) are available simultaneously (Fernandes and do Carmo-Silva, 2006). Fernandes and do Carmo-Silva (2006) compared the performance of the GPOLCA mechanism with that of the POLCA and MRP mechanisms using a multi-stage, multi-product production system under changing product mix and high variability as a test bed. The study showed that GPOLCA performed better by maintaining the same high level of throughput and a lower WIP.

OTHER HYBRID CONTROL MECHANISMS

The hybrid control mechanisms are obtained as a combination of other pull-control strategies. The extended kanban control mechanism (EKCM) and the generalized kanban control mechanism (GKCM) are well-known hybrid mechanisms. A brief introduction to hybrid mechanisms is presented.

The EKCM is a combination of the classical Kanban and base-stock mechanisms (Dallery and Liberopoulos 2000) and includes the two individual mechanisms as special cases. The mechanism depends on two parameters at every stage, namely the number of kanban cards at a stage and the base-stock level at each stage. It has to be noted that the number of kanban cards at a stage should at least be equal to the base-stock level at that stage. In this mechanism, a realization of demand causes an instantaneous replenishment order at all stages of the system as in a base-stock mechanism, and the kanban cards are not used to transmit demand to the preceding stages. The kanban cards are used for authorizing the transfer of parts to a downstream stage. Thus, the production capacity depends only on the number of kanban cards and not on the base-stock level.

Dallery and Liberopoulos (2000) described the workings of the EKCM and presented some of its properties. Chaouiya et al. (2000) extended the EKCM to an assembly system through the use of two variants of the EKCM. The variants differ only in the means by which the various sub-assemblies become available at the assembly station. Both variants require the presence of an assembly kanban to authorize a new subassembly. However, in the first variant, the components are released as soon as they become available, while they are made to wait for all components to be available in the second variant. The former is called the independent EKCM (IEKCM), and the latter is called synchronized EKCM (SEKCM). The study revealed that customer demands were satisfied earlier in the IEKCM.

The GKCM, proposed by Buzacott (1989), consists of the kanban, base-stock, and CONWIP mechanisms as its special cases (Frein et al., 1995). Like EKCM, it depends on two parameters at each stage – the number of kanban cards and the base-stock level. However, there is no restriction on the number of kanban cards to be at least equal to the base-stock level at that stage. This mechanism is equivalent to the Kanban mechanism when the base-stock level at stage *i*, S_i, equals the number of kanban cards at stage *i*, K_i, for all stages in the production system. When this situation occurs in a single-stage GKCM ($S = K$), the GKCM is equivalent to a CONWIP mechanism (Frein et al., 1995).

The GCKM consists of two synchronization stations at each stage. The first synchronization station at a stage consists of two queues that represent the finished part at that stage and the kanban from the downstream stage representing withdrawal authorization for a part from that stage to the downstream stage. The second synchronization station at that stage also consists of two queues representing the demand from the downstream stage for new parts at the stage and the kanban of that stage that authorizes demand information movement to the upstream stage (Duri et al., 2000).

Buzacott (1989) focused on developing queueing models for a tandem production system controlled by a GKCM where the kanban card was used for PA. However, the maximum stock held at a stage could differ from the number of kanban cards. Buzacott (1989) developed models for the single-stage and multi-stage production systems, where both kanban and MRP systems are special cases of the GKCM system. Frein et al. (1995) studied the influence of the design parameters (number of kanban cards and the base-stock level) on the efficiency of the GKCM. A GKCM was modeled as a multi-class CQN, where class *i* corresponded to stage *i* kanban cards. It has to be noted that all workstations and the synchronization station at the stage closed to external demand (stage 1) are visited by only one class of customers, while all other synchronization stations are visited by exactly two classes of customers. This network was solved using an approximation scheme that is similar in spirit to the one used by Di Mascolo et al. (1996). The study conducted on the single-stage GKCM revealed that the production capacity depended only on the number of kanban cards, while the stock-out rate depended mainly on the base-stock level. It was also shown that the multi-stage GKCM with planned inventory at stage 1 and *K* kanban cards at all stages is equivalent to a single-stage GKCM comprising all workstations with *K* kanban cards.

There are also other hybrid mechanisms that have been studied, like the CONWIP Kanban control (Bonvik et al. 1997) and extended CONWIP Kanban control (Boonlertvanich 2005).

COMPARISON OF CONTROL MECHANISMS

Spearman and Zazanis (1992) studied the behavior of the push and pull systems in an attempt to highlight the benefits of the Kanban and CONWIP systems. They showed that the pull system had less congestion due to the lower mean and variability of the WIP. The study also revealed that the CONWIP system outperformed the pure push systems and the pure pull (Kanban) system with respect to system throughput. It was also suggested with the possibility of WIP build-up in a CONWIP system, the Kanban system could perform better if the bottleneck was close to the end of the line.

Gstettner and Kuhn (1996) examined single-product flow lines with exponential processing times and unlimited demand at the final buffer with a focus on comparing the performance of the Kanban and CONWIP systems. The comparison revealed that the Kanban mechanism is more flexible than CONWIP because there are many ways of arranging the cards in a Kanban system, and by carefully selecting a card arrangement, one can easily achieve a given production rate at a lower WIP (Gstettner and Kuhn 1996).

Bonvik et al. (1997) used a four-workstation tandem production line under constant and varying demand rates as a test bed to compare the performance of the kanban, minimal blocking (another kanban model studied by researchers such as So and Pinault (1988), Mitra and Mitrani (1990), etc.), base-stock, CONWIP, and Kanban-CONWIP mechanisms. It was found that the performance of the base-stock and CONWIP mechanisms fell between that of the kanban and the other two mechanisms. CONWIP and Kanban-CONWIP mechanisms were found to be more robust with respect to the changes in the demand rate.

Duri et al. (2000) present a comparison of the three pull-control mechanisms – kanban, base-stock, and generalized kanban – for a three-stage production system. The study concluded that the generalized kanban which had the kanban mechanism and the base-stock mechanism as special cases was difficult to implement. It was also shown that the cost under these three mechanisms was the same when there were no delays in filling orders. Further, in the presence of delays in filling orders, the generalized kanban and base-stock mechanism were found to perform better.

SUMMARY

In the current competitive market scenario, companies need to ensure that customer orders are quickly satisfied with high-quality products at a low cost. Such a situation warrants a combination of the pull material control flow strategy (production is driven by demand realization) and the make-to-stock strategy (finished goods are kept in stock). The former strategy can aid in lowering WIP and manufacturing lead times. The latter strategy can increase the chance of instantaneous demand satisfaction. Several material control mechanisms have been employed to implement the pull material control flow strategy and an active area of research is in the development of new hybrid mechanisms that combine the desired features of some of the commonly used control mechanisms. Another area of research focuses on the development of analytical models to analyze the performance of production-inventory systems under various pull-control mechanisms.

This chapter has presented a detailed description of the commonly used pull-control mechanisms like kanban, base-stock, CONWIP, and POLCA and has touched upon a few hybrid mechanisms. It has also provided a review of the performance evaluation models that can aid in understanding the complex dynamics between system parameters in the various pull-control systems. The managerial insights and practical implications from these studies are useful for industry practitioners in deciding key business parameters. Additionally, the analysis of the applicability of different mechanisms for different production environments has been illustrated with a brief account of comparative studies found in the literature.

REFERENCES

Benjaafar, S., Cooper, W.L., and Kim, J.-S., On the benefits of pooling production-inventory systems, *Management Science*, 51(4), 548–565, 2005.

Benjaafar, S., Kim, J.-S., and Vishwanadham, N., On the effect of product variety in production-inventory systems, *Annals of Operations Research*, 126, 71–101, 2004.

Bhat, S., and Krishnamurthy, A., Production rate strategies for manufacturing systems with seasonal demands, *Proceedings of the 2012 Industrial and Systems Engineering Research Conference*, Orlando, FL, USA, 2012.

Bonvik, A.M., Couch, C.E., and Gershwin, S.B., A comparison of production-line control mechanisms, *International Journal of Production Research*, 35(3), 789–804, 1997.

Boonlertvanich, K., *Extended-CONWIP-Kanban System: Control and Performance Analysis*, Phd dissertation, Georgia Institute of Technology, Atlanta, GA, USA, 2005.

Buzacott, J.A., and Shanthikumar, J.G., *Stochastic Models of Manufacturing Systems*, Prentice Hall, Inc., Englewood Cliffs, NJ, USA, 1993.

Buzacott, J.A., Queueing models of kanban and MRP controlled production systems, *Engineering Costs and Production Economics*, 17, 3–20, 1989.

Chaouiya, C., Liberopoulos, G., and Dallery, Y., The extended kanban control system for production coordination of assembly manufacturing systems, *IIE Transactions*, 32(10), 999–1012, 2000.

Clark, A.J., and Scarf, H., Optimal policies for the multi-echelon inventory problem, *Management Science*, 6(4), 475–490, 1960.

Dallery, Y., and Liberopoulos, G., Extended kanban control system: Combining kanban and base stock, *IIE Transactions*, 32(4), 369–386, 2000.

Deleersnyder, J.-L., Hodgson, T.J., Muller, H., and O'Grady, P.J., Kanban controlled pull systems: An analytic approach, *Management Science*, 35(9), 1079–1091, 1989.

Di Mascolo, M., Frein, Y., and Dallery, Y., An analytical method for performance evaluation of kanban controlled production systems, *Operations Research*, 44(1), 50–64, 1996.

Duenyas, I., and Hopp, W.J., Estimating the throughput of an exponential CONWIP assembly system, *Queueing Systems*, 14, 135–157, 1993.

Duenyas, I., and Hopp, W.J., Estimating variance of output from cyclic exponential queueing systems, *Queueing Systems*, 7(3), 337–353, 1990.

Duri, C., Frein, Y., and Di Mascolo, M., Comparison among three pull-control policies: Kanban, base stock, and generalized kanban, *Annals of Operations Research*, 93, 41–69, 2000.

Fernandes, N.O., and do Carmo-Silva, S., Generic POLCA – A production and material flow control mechanism for quick response manufacturing, *International Journal of Production Economics*, 104, 74–84, 2006.

Frein, Y., Di Mascolo, M., and Dallery, Y., On the design of generalized kanban control systems, *International Journal of Operations and Production Management*, 15(9), 158–184, 1995.

Gavish, B., and Graves, S.C., A one-product production/inventory problem under continuous review policy, *Operations Research*, 28(5), 1228–1236, 1980.

Gavish, B., and Graves, S.C., Production/inventory systems with a stochastic production rate under a continuous review policy, *Computers & Operations Research*, 8(3), 169–183, 1981.

Graves, S.C., and Keilson, J., The compensation method applied to a one-product production/inventory problem, *Mathematics of Operations Research*, 6(2), 246–262, 1981.

Gstettner, S., and Kuhn, H., Analysis of production control systems kanban and CONWIP, *International Journal of Production Research*, 34(11), 3253–3273, 1996.

Hopp, W.J., and Spearman, M.L., Throughput of a constant work in process manufacturing line subject to failures, *International Journal of Production Research*, 29(3), 635–655, 1991.

Huang, C., and Kusiak, A., Overview of kanban system, *International Journal of Computer Integrated Manufacturing*, 9(3), 169–189, 1996.

Huang, P.Y., Rees, L.P., and Taylor III, B.W., A simulation analysis of the Japanese just-in-time technique (with Kanbans) for a multiline, multistage production system, *Decision Sciences*, 14(3), 326–344, 1983.

Jain, S., and Raghavan, N.R.S., Analysis of base-stock controlled production-inventory system using discrete-time queueing models, *IEEE International Conference on Automation Science and Engineering*, pp. 37–42, Edmonton, Canada, Aug 2005.

Kamath, M., and Srivathsan, S., A comparative evaluation of analytical approximations for production-inventory networks, *Computers & Industrial Engineering*, 62, 644–652, 2012.

Karaesman, F., Liberopoulos, G., and Dallery, Y., Production/inventory control with advance demand information, in *Stochastic Modeling & Optimization of Manufacturing Systems & Supply Chains, Vol. 63: Int. Series in Operations Research & Management Science*, Shanthikumar, J.G., Yao, D.D., Zijm, W.H.M., Hillier, F.S., and Price, C.C., Eds, Springer US, 243–270, 2003.

Karaesmen, F., Buzacott, J.A., and Dallery, Y., Integrating advanced order information in make-to-stock production systems, *IIE Transactions*, 34, 649–662, 2002.

Karaesmen, F., Liberopoulos, G., and Dallery, Y., The value of advance demand information in production/inventory systems, *Annals of Operations Research*, 126, 135–157, 2004.

Karmarkar, U.S., and Kekre, S., Batching policy in kanban systems, *Journal of Manufacturing Systems*, 8(4), 317–328, 1989.

Krajewski, L.J., King, B.E., Ritzman, L.P., and Wong, D.S., Kanban, MRP, and shaping the manufacturing environment, *Management Science*, 33(1), 39–57, 1987.

Krishnamurthy, A., and Ge, D., Integrating kanban control with advance demand information: Insights from an analytical model, in *Rapid Modelling for Increasing Competitiveness*, Reiner, G., Ed., Springer London, 131–141, 2009.

Krishnamurthy, A., and Suri, R., Performance analysis of single stage kanban controlled production systems using parametric decomposition. *Queueing Systems*, 54, 141–162, 2006.

Krishnamurthy, A., and Suri R., Planning and implementing POLCA: A card-based control system for high variety or custom engineered products. *Production Planning and Control*, 20(7), 596–610, 2009.

Krishnamurthy, A., *Analytical performance models for material control strategies in manufacturing systems*, PhD dissertation, University of Wisconsin-Madison, Madison, WI, USA, 2002.

Lee, Y.-J., and Zipkin, P.H., Tandem queues with planned inventories, *Operations Research*, 40(5), 936–947, 1992.

Leung, Y.T., and Suri, R., Performance analysis of discrete manufacturing systems, *IEEE Control Systems Magazine*, 77–86, 1990.

Liu, L., Liu, X., and Yao, D.D., Analysis and optimization of a multistage inventory-queue system. *Management Science*, 50(3), 365–380, 2004.

Mitra, D., and Mitrani, I., Analysis of a kanban discipline for cell coordination in production lines. I. *Management Science*, 36(12), 1548–1566, 1990.

Mitra, D., and Mitrani, I., Analysis of a kanban discipline for cell coordination in production lines, II: Stochastic demands. *Operations Research*, 39(5), 807–823, 1991.

Neuts, M.F., Matrix-geometric solutions in stochastic models: An algorithmic approach, *Johns Hopkins Series in the Mathematical Sciences, vol. 2*, The Johns Hopkins University Press, 1981.

Ohno, T., *The Toyota Production System: Beyond Large Scale Production*, Productivity Press, Portland, OR, USA, 1988.

Riezebos, J., Design of POLCA material control systems, *International Journal of Production Research*, 48(5), 1455–1477, 2010.

Sarker, B.R., and Fitzsimmons, J.A., The performance of push and pull systems: a simulation and comparative study, *International Journal of Production Research*, 27(10), 1715–1731, 1989.

Satyam, K., and Krishnamurthy, A., Performance evaluation of a multi-product system under CONWIP control, *IIE Transactions*, 40(3), 252–264, 2008.

Sivaramakrishnan, S., and Kamath, M., Analytical models for multi-stage make-to-stock systems, *Proceedings of the 6th Industrial Engineering Research Conference*, pp. 795–800, Norcross, GA, USA, 1996.

Sivaramakrishnan, S., *Analytical Models for Performance Evaluation of Production-Inventory Systems*, Phd dissertation, Oklahoma State University, Stillwater, OK, USA, 1998.

So, K.C., and Pinault, S.C., Allocating buffer storages in a pull system, *International Journal of Production Research*, 15(12), 1959–1980, 1988.

Spearman, M.L., Hopp, W.J., and Woodruff, D.L., A hierarchical control architecture for Constant Work-in Process (CONWIP) production systems, *Journal of Manufacturing and Operations Management*, 2, 147–171, 1989.

Spearman, M.L., Woodruff, D.L., and Hopp, W.J., CONWIP: A pull alternative to kanban, *International Journal of Production Research*, 28(5), 879–894, 1990.

Spearman, M.L., and Zazanis, M.A., Push and pull production systems: Issues and comparisons, *Operations Research*, 40(3), 521–532, 1992.

Srivathsan, S., *Analytical Performance Modeling of Supply Chain Networks*, Master's thesis, Oklahoma State University, Stillwater, OK, USA, December 2004.

Srivathsan, S., *Modeling Inventory Information Visibility in Supply Chain Networks*, PhD dissertation, Oklahoma State University, Stillwater, OK, USA, 2012.

Srivathsan, S., and Kamath, M., An analytical performance modeling approach for supply chain networks, *IEEE Transactions on Automation Science and Engineering*, 9(2), 265–275, 2012.

Sugimori, Y., Kusunoki, K., Cho, F., and Uchikawa, S., Toyota Production System and kanban system materialization of just-in-time and respect-for-human system, *International Journal of Production Research*, 15(6), 553–564, 1977.

Suri, R., Sanders, J.L., and Kamath, M., Performance evaluation of production networks, in *Handbooks in Operations Research and Management Science, Vol. 4: Logistics of Production and Inventory*, Graves, S.C., Rinnooy Kan, A., and Zipkin, P., Eds., Elsevier Science Publishers, B.V., 199–286, 1993.

Suri, R., *Quick Response Manufacturing: A Companywide Approach to Reducing Lead Times*. Productivity Press, Portland, OR, USA, 1998.

Svoronos, A., and Zipkin, P.H., Evaluation of one-for-one replenishment policies for multi-echelon inventory systems, *Management Science*, 37(1), 68–83, 1991.

Vandaele, N., Nieuwenhuyse, I.V., Claerhout, D., and Cremmery, R., Load-based POLCA: Integrated material control system for multiproduct, multimachine job shops, *Manufacturing and Service Operations Management*, 10(2), 181–197, 2008.

Whitt, W., The queueing network analyzer, *The Bell System Technical Journal*, 62(9), 2779–2815, 1983.

Whitt, W., Towards better multi-class parametric-decomposition approximations for open queueing networks, *Annals of Operations Research*, 48, 221–248, 1994.

Zipkin, P.H., Processing networks with planned inventories: Tandem queues with feedback, *European Journal of Operational Research*, 80(2), 344–349, 1995.

Srivathsan, Sandeep and Manjunath Kamath (2014), "Modeling Production-Inventory Systems," Chapter in *Handbook of Industrial & Systems Engineering*, Second Edition, Taylor & Francis Group, CRC Press, Boca Raton, FL.

6 Time as a Critical Factor of Production

TIME AND PRODUCTION

Time is a critical factor in the adoption and utilization of technology. If we worship time, we will respect the timely impact of technology. Faluyi (2006) presents time as a factor of production. In a similar vein, Badiru (2006) presents a poem on the passage of time. Benjamin Franklin, so long ago, presented a parody of time as a stuff of life.

> Dost thou love life? Then do not squander time, for that is the stuff life is made of
>
> **– Benjamin Franklin**

POEM OF TIME (BADIRU, 2006)

On The Flight of Time.
What is the speed and direction of Time?
Time flies, but it has no wings.
Time goes fast, but it has no speed.
Where has time gone? But it has no destination.
Time goes here and there, but it has no direction.
Time has no embodiment. It neither flies, walks, nor goes anywhere.
Yet, the passage of time is constant.

– Adedeji Badiru

Faluyi (2006) uses an intriguing Nigerian storytelling style to convey the sense of the importance of time in production activities. He recounts: The young Nigerian had just arrived in Cork, Southern Ireland (Eire), from across the channel. After the usual pleasantries with the gentleman who came to meet him, the visitor asked how soon the day would start on the following day. With a glint in his eye, the Irish man replied, "There is an old Irish saying that when God made time, he made plenty of it." "So relax my friend, you are on this side of the channel now." The point being made was not lost on the visitor. Why, only sometime earlier the Duke of Edinburgh was visiting a city in the British Midlands. The City Elders were at the railway station to welcome him. As the Duke alighted from the train, he went right up to the elders to apologize for coming late. Yes, his train had arrived two minutes late! That was the time when British Railways was under fire for not keeping to its schedule.

Imagine it, the Royal Train operated by British Railways arrived two minutes late. Such was the British attitude to timekeeping and lateness.

DOI: 10.1201/9781032659220-6

The Irish are also right. No one can contest that when God made time, he made plenty of it. As the God of Eternity, he is not subject to time constraints. However, the authentic account of creation shows that when He created time, He introduced seasons into it. In fact, the account of creation states "And God went on to say let luminaries come to be in the expanse of the heavens to make division between the day and the night and they must serve as signs and for seasons and for days and years" [Gen. 2:14]. So, from the beginning, the Creator intended that time should be measured, it was not fluid and unquantifiable. There were to be seasons, days, and years.

Rural communities all over the world appreciate that. There is a season for planting and a season for harvesting. Farmers know it and abide by it. No farmer plants and harvests with immediate effect. The farmer has a wholesome respect for time. For him, time is a factor of production, but the scale for measurement is much larger and has room for flexibility. A planting season or harvesting season does not have narrow precise and rigid timing limits for planting and harvesting. They have no need for hours and minutes.

ORIGIN OF HOURS AND MINUTES

In the earliest of times, the Egyptians sought accuracy with respect to the use of time. First, the calendar year was established and then divided into 36 decans, each ten days long plus five extra days for a 365-day year. Each decan corresponded to a third of a zodiacal sign and was represented by a decanal constellation.

The Egyptians and later the Babylonians (about 300–100 B.C.E) decided to divide the day from sunrise to sunset into 12 parts that are called hours. They also decided to divide the night, the time from sunset to sunrise, into 12 hrs. But daylight and the night are not of the same length all year round. This system of measuring time was not very accurate because the actual length of the hour changed depending on the time of year. Someone finally figured out that by dividing the whole day into 24 hrs (12 hrs of the day plus 12 hrs of the night), the time could be measured much more accurately. This system is now universally in use. This point is lost on Nigerians. We still consciously think of time in two divisions of 12 hrs each and not a day of 24 hrs. Thus, we speak of 6.00 hrs in one division as 6.00 a.m. and 6 hrs in the second division as 6.00 p.m. Some other community speaks of 6.00 and 18.00 hrs. Certainly, that conveys more accurately how much time of the day has been spent.

In spite of this very early consciousness of time, time was not seriously considered as a factor of production. In calculating costs of production, time was not taken into consideration because no monetary value was ascribed to time.

In rural communities, north and south of the equator, time is seen largely in terms of seasons, namely spring, summer, autumn, and winter. It was a long enough time to permit flexibility. The sub-division into days, hours, minutes, and seconds had no place. It was within that environment that the Industrial Revolution started in Europe. Economists and managers saw production only in terms of the material contributions to production, namely men, machines, money, and later, management. Time was no factor of production then.

It was not until the time of Fredrick Taylor, early in the 20th century, that time took its rightful place among the factors of production. Taylor, who was reputed as one of the fathers of industrial engineering, set out to study work as a subject. He sought to break every activity into its component parts, and it is the aggregate of those components that becomes the activity. That was the beginning of work study, which consists of time and motion study, a key aspect of any industrial engineering curriculum. Every motion took a precise amount of time, which can be measured and given a precise value. So, time now has a value, which can be managed. Today, man has moved on to study and research not only materials management but also time management. Time has come to be recognized as a factor of production, which must be properly managed. Amen!

Labor then has an hourly rate of earning, and payments were made in accordance with this agreed hourly rate of earning both for manual skills and professional skills. A consulting engineer would be paid in accordance with the hours he spends on the assignment, taking into consideration the level of proficiency required for the assignment. So would an accountant, as would a plumber, a motor mechanic, or an electrician. Each would be paid in accordance with the hourly rate assigned to the complexity of his activity.

Interestingly, as the hourly rate for craftsmen rose steadily in Europe, the citizenry became restive and an increasing number of them sought to do the simpler craft tasks themselves. For example, a plumber was no longer called in to unblock kitchen sinks and later electricians were not invited to change fuses and light bulbs even on appliances. An age of do it yourself (DIY) was born. Today, DIY has grown into a culture of maintenance that is sustained by specialized small tools and a regime of books and other materials that are specially provided for the untrained DIY enthusiast. Nowadays, online YouTube guides are increasingly replacing printed DIY guidebooks. And that truly is what is required – enthusiasm. We talk glibly about a lack of maintenance culture in Nigeria. But truly, what has been done to promote such a culture? The environment that sustains such a culture is just not there. Who is there who wants to spare his time to have his equipment or machine opened up for diagnosis and repairs? Rather, the vast majority wants it replaced, and we may as well start to accept that replacement of machines or machine components is an acceptable mode of maintenance. There is a place for replacement maintenance within the context of maintenance management. And for us, the literature and other materials, even spare parts, are not available in Nigeria to facilitate repairs. So, we need to see a maintenance culture in another context, not necessarily a diagnosis of faults and repairs. All of these have direct and serious implications for workforce development and retention.

THE NIGERIAN TIME

In the context of workforce utilization, there is a pervasive Nigerian time. From the earliest of time, Nigerians have not seen time in terms of its smaller units of hours, minutes, and seconds. For the Nigerian workforce, the Irish saying is: "when God made time he made plenty of it." So, why the constraint of hours and minutes? So, when a Nigerian says, "I'll see you tomorrow," he (or she) means, "I'll see you sometime during the daylight hours tomorrow." If he is pressed for a more definite time, then he says in the morning or in the afternoon. And that could mean any time before noon or the time the sun is at its highest point or any time after that before it goes completely dark. Time for the Nigerian workforce could be quite fluid.

CULTURAL DIVIDE ABOUT TIME

There is this anecdote that clearly illustrates this point. In the mid-1930s, the Resident of Oyo (an official of the British Government) administering the then Oyo province in today's south-western Nigeria had an appointment to see the Alaafin of Oyo in the Resident's office at 10:00 a.m. Indeed, a few minutes before 10:00 a.m., the Resident was at the door, waiting to receive a highly respected and very senior ruler in Yorubaland. The Resident waited and waited. The Alaafin arrived with his retinue of drummers, praise singers, and court jesters at about 12 noon – two hours behind schedule, late! The Resident was furious and he did little to hide his displeasure at what he considered uncivil and indeed bad manners. Imagine keeping the King of England's representative waiting for two hours! The Alaafin was told that the Resident didn't take kindly to what he did; the appointment was for 10:00 a.m. and to see what time the Alaafin was coming in, and with such pomp and pageantry. The Alaafin calmly told the interpreter to tell the Resident that it is when the Alaafin arrives that it is 10:00 a.m. and not before. Yes, man is not bound by hours and minutes. It is what man calls it that it is. We are back at it again, when God made time, he made plenty of it.

Expatriates, who failed to understand this mental attitude of Nigerians in apportioning time came to describe this approach as "African time." No, Africans don't have a time of their own. It is just that Africans, nay; the Nigerians do not operate within the precise context of hours, minutes, and seconds. Even at this time of Nigerian history, how many meetings and events start at the scheduled time? Some people have sought to call attention to this failing. When they send out an invitation to a meeting or an event, they add the word prompt at the end of the time. Such as the 10:00 a.m. prompt. We have not come to understand what the word "prompt" means as the meeting or the event will start when it starts without any reference to the scheduled time, prompt, or no prompt. The military added their own dimension to the concept of time in Nigeria. It was the military that introduced "with immediate effect" or "Effective immediately." For them, time has no impact on tasks. How, indeed, can anything be done with immediate effect? An official who is retired with immediate effect is required to write handing over notes! This requires some time to accomplish. A bridge (or was it a girder?) was to be completed "with immediate effect." Certainly, some actions are involved, and these actions involve some motions, which would take time. So, what really then is the meaning of "with immediate effect"? That's an arrant disrespect for time.

THE FALLACY OF PROJECT COMPLETION TIME

An example of this disrespect for time, with an unpleasant result, is shown in the construction of culverts and slabs. It is the norm to construct these culverts and slabs with concrete, sometimes reinforced concrete. Ordinarily, concrete takes some 28 days to cure. In other words, it will take some 28 days for the mixture of stones, cement, sand, and water to acquire the strength and characteristics of concrete. But not so in Nigeria; the country cannot wait. We have no time. So, after a few days, the culvert is put into use, or the slab is built upon. We cannot wait. Time has no value. The culvert fails or the slab collapses. We need to accept that time is a factor of production. Some time is required to carry out some tasks effectively, efficiently, and safely.

There is yet another angle to this indifferent attitude toward time. It is costing. A plumber or an electrician doing exactly the same job in Ikoyi or Victoria Island would charge differently from what he would charge in Mushin or Ajegunle. Why? The craftsman does not have an hourly rate value. His charge is based on his client's ability to pay. And that's another twist. Sometimes, the craftsman may have to source for materials and spare parts. He may have to go and find them and then come to collect money to buy them. Who pays for the time spent on that effort? The client only takes into consideration just the time spent on the assignment. So, how accurate is the contribution of this informal sector to the nation's gross national product?

BUDGET CYCLE CONUNDRUM

Let us look at it at another level. The national budget and the state government budgets are usually prepared for one year, namely from January to December. In other words, the government has determined what work it wants to do within the next 12 months and has put a monetary value to it. However, it is sad that what is often of concern to most commentators, including government officials, who are involved in the preparation and approval of the budget, is the amount of money allocated and spent, not the actual work done. Of course, no one cares about the time required to finish the work assigned. It is clear that time is not a relevant factor. Otherwise, how can one explain the passing of the appropriation bill and signing it into law in March for a budget that would expire in December? So, the budget approval is to execute works scheduled for 12 months in nine months, at best. So, how much does the government appreciate time as a factor of production?

No, Nigerians' appreciation of time must change. And the change is basic and must be at all levels. It seems good that current reform programs should be underpinned by a major shift in the attitude toward time.

NATIONAL TIME STANDARDS

A symbolic gesture may well be used to kick off the campaign to promote a change of attitude toward time. There are 24 hrs in a day, but we treat it as 12 hrs in that the measurement of time is within a 12 hr framework. The counting starts at 12 midnight and when it gets to noon we start counting from zero again. The Federal Government, by fiat or by an Act of the National Assembly, if that is the appropriate thing to do, should make our counting of time to be continuous from 12 midnight to 12 midnight. So, 1 p.m. that is 1 hr after noon becomes 13:00 hrs, and 5 p.m. becomes 17:00 hrs, and so forth. That is, indeed, the military time system. It will impress on all and sundry that we have spent 17 hrs out of the 24 hrs we have in a day – not an illusory 5 hrs. It is important that everyone should be conscious of the fact that time is a consumable product. Only 24 hrs are available in a day. It cannot be extended no matter how much one wishes or tries. So, a continuous reckoning of time may drive home that point, subconsciously, but effectively. Comically, Badiru (2013, 2020, 2022) presents the secrets of the 25th hr of the day. This is not from a time-creation fallacy but from a time-saving and conservation approach.

GETTING A SHIFT ON TIMELINE

The government, by its pronouncements and actions, should also acknowledge that time is finite and show proper respect for it. So, a 12-month budget that is for January to December should not be signed into law in March or whenever. The annual budget should be debated, approved, and signed into law not later than the 31st day of December of the preceding year. Thus, there will truly be an implementation period of 12 months for the approved budget. In that circumstance, budgeting will make more sense. Those planning the budget will include in irony what they have the capacity to do within 12 months. Performance appraisal will be more realistic and meaningful. And how nice it would be when performance is appraised not in terms of money spent but in terms of actual physical achievement.

Also in giving assignments, realism should be considered in allocating time. If you ask for a thorough investigation of a major accident or a devastating disturbance and you ask the panel to submit its report within two weeks, then you are saying you do not want a thorough investigation or a good report. These things, if they are going to be acceptable, take time and the government must recognize that. Ridiculous timelines give the impression that the authority concerned is not aware of the volume of work involved or is not really concerned about doing a good job.

A consulting consortium was, once, requested to plan and design a Medical Diagnostic Center within three months. When the professionals protested that the time allowed was too short, they were asked if they wanted the job. "I suppose you are very busy now and you can't take the job," it was suggested to them. The planning and design was completed in four months or so. But the Diagnostic Center was never built and that was over 25 years ago. Some readers here, probably, have similar experiences. The government appears to think that money and manpower can shorten time. Not so! A popular parody in project management is to put more men on a job to get the job done.

This brings up a popular anecdote, in this regard, from that period of time when money seemed to be all that mattered. A renowned architect used to visit his construction sites on Saturdays. On many of those occasions, he took his son, John along. Each week John listened to his father admonish the contractor that the job was slow and dragging. "Put more men on the job" daddy would counsel. About this time, John's mother was pregnant. John saw mummy's tummy shooting out, and he was told that mummy was making a baby inside her tummy. John waited for months to see mummy produce the baby she was making, and nothing was forthcoming. So, John approached his father and said "Daddy, why don't you admonish mummy to put more men on the job and accomplish this work of making a baby in time." Can putting more men on the job speed up the making of a baby? Time is a factor of production, and it will be good for public officials to remember that when they are awarding construction contracts, in particular, and, indeed, any contract, a wholesome respect for time be considered as essential for the productive activity. "With immediate effect" is not a viable option for getting good results. As Kenneth Patton puts it "By labour, we can find food and water, but all of our labour will not find for us another hour."

TIME EFFECT ON PRODUCTIVITY

Nigerians appear to be more concerned about the appearance of being busy rather than the substance of what is being achieved. Consider this example: A project has been identified. Promptly, a committee is set up and, of course, meetings are arranged. The public service is well known for this. Senior officers go from one meeting to another meeting all day. They are very busy, attending meetings. When then is the work done? Time is not often allowed for the implementation of the decisions made at such meetings. Time is not set aside for reviews of what is being done or for reflections on the decisions required and made. Consequently, time is not spent efficiently or effectively. Productivity is at its low ebb in such cases. This brings up Badiru's treatise of management meetings (Badiru, 1996), which is echoed below.

PERSPECTIVES ON WORKFORCE MEETINGS

Parkinson's law says, "Work expands to fill the available time."

Deji Badiru's mutation says, "Talk expands to fill the available time."

Meetings represent one avenue for information flow and knowledge sharing for group decision-making. Effective management of meetings is an important skill for any administrator or manager. Unfortunately, meetings often degenerate into time-wasters; consequently, obstructing productivity and detracting from other essential functions. This is because most meetings are poorly organized, improperly managed, called at the wrong time, or even unnecessary. In some organizations, meetings are conducted as a matter of routine requirement rather than necessity. While meetings are essential for communication and decision-making, they accomplish nothing if not managed properly. A meeting of 30 people wasting only 30 minutes in effect wastes 15 full hrs of employee time. That much time, in any organization, amounts to thousands of dollars in lost time. It does not make sense to use a one-hour meeting to discuss a task that will take only five minutes to perform. This is analogous to hiring someone at $100,000 annual salary to manage an annual budget of $5,000. In 1993, a newspaper columnist/humorist, Dan L. Stewart, wrote about management meetings:

> Management meetings are rapidly becoming this country's biggest growth industry. As nearly as I can determine, the working day of a typical middle manager consists of seven hours of meetings, plus lunch. Half a dozen years ago at my newspaper, we hired a new middle management editor with an impressive reputation. Unfortunately, I haven't met her yet. On her first day at work, she went into a meeting and has never come out.
>
> Stewart concludes his satire with "I'm expected to attend the next meeting. I'm not sure when it's scheduled exactly. I think they're having a meeting this afternoon about that."

In the past, when an employee had a request, he went to his boss, who would say "yes" or "no" right away. The whole process might take less than one minute out of the employee's day. Nowadays, several hierarchies of meetings may be needed to review the request. Thus, we may have a departmental meeting, a middle management staff meeting, an upper management meeting, an executive meeting, a steering committee meeting, ad hoc committee meeting, an off-site retreat, and a meeting with outside consultants; all for the purpose of reviewing the simple request. Deji's perspectives about meetings are summarized below:

1. Don't call for a meeting, unless it is absolutely needed.
2. Most of the information passed out at meetings can be more effectively disseminated through simple memos. Emails should be fully exploited to replace most meetings.
3. The point of diminishing returns for any meeting is determined by the number of people that are actually needed for the meeting. The larger the number of people at a meeting, the lower the meeting's productivity. The extra attendees only serve to generate unconstructive and conflicting ideas that only impede progress.
4. Not being invited to a meeting should be viewed as an indication of the high value placed on an individual's time within the organization.
5. Regularly scheduled meetings with specific time slots often become a forum for social assemblies if not managed properly.
6. The satirical adjourn time of a meeting is equal to the scheduled start time plus five times the number of agenda items minus the actual start time. Mathematically, this is expressed as:

$$L = (T + 5N) - S$$

where:

L = optimal length
T = scheduled time
N = number of agenda items
S = meeting start-up time (i.e. actual call-to-order time)
$\mathbf{S \leq 5N}$

The subtraction of S in the expression is to encourage calling the meeting to order promptly. The larger the value of S, the less time is available for actual business. If **S = 0**, then the total meeting time is allocated for conducting business. If **S = 5N**, then there will be no time left to hold the meeting. Because it is difficult to do away with all meetings, we must maximize the output of those that remain by using these guidelines:

1. Do pre-meeting homework.
   a. list topics to be discussed (agenda).
   b. establish the desired outcome for each topic.
   c. determine how the outcome will be verified.

   d. determine who really needs to attend the meeting.
   e. evaluate the suitability of meeting time and venue.
   f. categorize meeting topics (e.g., announcements, important, and urgent).
   g. assign time duration to each topic.
   h. verify that the meeting is really needed.
   i. consider alternatives to the meeting (e.g., memo, telephone, and electronic mail).
2. Circulate a written agenda prior to the meeting.
3. Start the meeting on time.
4. Review the agenda at the beginning.
5. Get everyone involved; if necessary, employ direct questions and eye contact.
6. Keep to the agenda; do not add new items unless absolutely essential.
7. Be a facilitator for meeting discussions.
8. Quickly terminate conflicts that develop from routine discussions.
9. Re-direct irrelevant discussions back to the topic of the meeting.
10. Retain leadership and control of the meeting.
11. Recap the accomplishments of each topic before going to the next. Let those who have made commitments (e.g. promise to look into certain issues) know what is expected of them.
12. Have empathy for those invited to meetings and end the meetings on time.

Regardless of the structure and expediency exercised in conducting meetings, the real work is done behind the scenes. Note that, as social animals, we sometimes need separate "social meetings" where we can have all the chips and drinks we want and take all the time that we want.

By the way, one definition of a meeting is "the confusion of one man, multiplied by the number of people present at the meeting," in accordance with the formula promulgated above.

In many countries, the concept of productivity is at the heart of the management of resources. Not so in Nigeria. Here, it is enough that a task has been done, no one cares for how long it takes or how well it is done. Perhaps an hourly rate of payment will call attention to this. If a member of the industrial workforce were to know that his (or her) pay depends on how much he can achieve in the time allowed for him, perhaps, he would be more particular as to how he does his (or her) job. Wage negotiations would be more meaningful. Productivity can be measured and appropriately rewarded.

RIGHTSIZING FOR TIME FACTOR

Currently, the talk is about the rightsizing of many public organizations. While some of these organizations maintain idle staff members, some staff members, in the same organization, are overworked and do a lot of genuine overtime work. Overtime should not be viewed as a part of a normal day's work. Dr. Donald Wetmore observes that life consists of seven vital areas: Health, family, financial, intellectual, social, professional, and spiritual. While it is not possible to spend time every day in each area, it is, however, important that, in the long run, spending a sufficient quantity and quality of time in each area would produce a balanced life. For example, failure to take time now for health will, most likely, make one to take time for illness later, or ignoring the family now could cost a lot of time to re-establish relationships later, where the family ties break up. Overtime or overwork arrangements disrupt the balance in a person's life, and it is not productive. It is a sure source of stress.

CONSTRUCTIVE TIME MANAGEMENT

The life balancing mentioned in the preceding section brings up another work-balancing template offered by Badiru (2013), regarding dividing each 24-hour day into three segments of regular work, home life, and sleep. Efficient time management is important for both corporate and personal success. One simple essence of time management at the corporate level is starting meetings at the scheduled time. Rarely do meetings start on time within the Nigerian context. But to be fair, this is also true in many other nations, even in some advanced economies. Everyone has come to accept that the meeting would not start on time. So, no one plans to be at the venue on time. They set out for the venue late and then blame the traffic for coming late. If all were time conscious, less time would be lost at the commencement or start of an event. If 20 people were at a meeting or an event that started 15 minutes late, a total of five productive hours have been lost. If it started 30 minutes late, a total of ten productive hours have been lost. If, indeed, we have an average hourly rate for those present, the value of the time lost, just for waiting to start, will be seen for what it truly is, a colossal waste of human resources.

If more people were to use the hourly rate for calculating their income, how much money is lost to traffic jams will have a greater impact. More often than not, these traffic jams are the result of indiscipline. A driver or two cannot wait, time must be abridged, or even eliminated because each driver has something that is considered very important to do. So, drivers jump queues or seek to join the traffic at odd and dangerous points, thus making a complete mess of the traffic, which could take hours to clear up. A healthy respect for time and its value would prevent traffic mishaps, as all will keep in mind the cost of the end result of a traffic jam.

PERSONAL TIME VERSUS CORPORATE TIME

At the personal level, time management is more challenging. It is a type of self-management. Different people may judge identical lengths of time quite differently. Time can "fly"; that is, a long period of time can seem to go by very quickly or can seem to drag on, as in when one performs a boring task. In explaining his theory of relativity, Albert Einstein has been quoted as saying that although sitting next to a pretty girl for an hour feels like a minute, placing one's hand on a hot stove for a minute feels like an hour. Consequently, at the personal level, time management requires an important shift in focus. Many people spend their days in a frenzy of activity but achieve very little because they are not concentrating on the right things.

There is a need to concentrate on results and not on being busy. Goals must be precisely set, priorities established, and planning of time written out. Such careful self-management would produce an efficient and effective use of time.

Yes, time is an important factor in production. Several books have been written on it, and courses and seminars have been delivered on time management. Computer software tools are even available to satisfy the need to control employees' time, make it easier to coordinate work, and increase accountability of the workforce.

With positive changes in the management of time, the workforce will come to appreciate time as a finite factor of production.

REFERENCES

Badiru, Adedeji B. (1996), *"Perspectives on Meetings,"* Essay on Management Practice, School of Industrial Engineering, University of Oklahoma, Norman, OK.

Badiru, Adedeji B. (2006), "The Flight of Time – What is the speed and direction of Time?" personal collection of life's poems, Knoxville, TN, USA.

Badiru, Deji (2013), *8-by-3 Paradigm for Time Management: Balancing Work, Home, and Leisure*, iUniverse, Bloomington, Indiana.

Badiru, Deji (2020), *The Twenty-Fifth Hour: Secrets to Getting More Done Every Day*, iUniverse, Bloomington, Indiana.

Badiru, Deji (2022), *Rapidity: Time Management on the Dot*, iUniverse, Bloomington, Indiana.

Faluyi, A. O. (2006), *Time: A Factor of Production*, Academy Lecture, The Nigerian Academy of Engineering, University of Lagos, Lagos, Nigeria, March 16, 2006.

7 Technology and Innovation Adoption Template

TECHNOLOGICAL SPREAD

The technology and innovation race is on, and it is gaining speed. The technology adoption timeline is a critical factor for success. Technology is all around us, in verbiage, promulgations, and actions, particularly in the prevailing age of AI. Everywhere we turn these days, everyone is proclaiming embracing AI technology as a mark of the pursuit of innovation. All kinds of progress are showcased with innovation, such as, "scaling up innovation," "exercising innovation," and "more/significant/faster innovation," just to name a few. There is a plethora of innovation symposia, fora, conventions, conferences, seminars, and meetings held all over the world. Similarly, innovation *corridors*, *centers*, *institutes*, and *platforms* have emerged in recent years. Even occupational job series are emerging with titles such as "Innovation Coordinator," "Innovation Facilitator," and "Innovation Accelerator." The website, https://www.etymonline.com/word/innovation, presents an interesting view of innovation, what it is, and what it portends (good and bad). The mere use of the word has exploded exponentially over the last 70 years. It appears that innovation has emerged as a free-for-all pursuit. Simply put, etymology of innovation implies "to change; to renew." A meaning from the 1540s indicates innovation as "a novel change, experimental variation, new thing introduced in an established arrangement." Innovation carries an element of being life-changing; in other words, a tremendous contribution to the quality of life such that things will never be the same. This thought alone evokes a great deal of excitement. Engineers, scientists, inventors, analysts, and leaders reading this book will probably feel a rush of energy from contemplating the potential of their endeavors when innovation makes its way into the process. This is probably best summed up by this quote:

> While necessity is the mother of invention, creativity is the father of innovation.
>
> **– Melinda L. Tourangeau**

Our willingness to create, to test the art of the possible, to take risks, to experiment, and to take pleasure in our achievements fuels the flames of creativity to spark innovation. An invention remains merely a great idea until it is transformed into an innovation. When innovation arrives, the world is forever changed. That will be just one of the metrics proposed in this book. Are you ready to change the world? Keep reading…

DOI: 10.1201/9781032659220-7

In all of this scrambling about innovation, one thing that is consistently missing is the measurement and affirmation of innovation. Is it sufficient to merely incarnate the word innovation and expect the world to believe? Of course, not. Anchoring innovation in metrics, rubrics, and standards can benefit both progenitors and society. This is the genesis of this book. Innovation conjures the presence of a new method, idea, and product. Innovation metrics are therefore essential. How do we know when we have achieved innovation, granted that innovation is essential for the ongoing success of any organization?

IMPORTANCE OF MEASUREMENTS

After all the preceding discussions, the questions that come to mind include the following:

- How do we measure innovation on a scale?
- How do we affirm each claim of innovation?
- Does innovation always connote the same thing across every spectrum?

If something has not been (or cannot be) measured and affirmed, its authenticity and value will always be debatable. For this reason, this book intends to set out a solid framework for measuring and "grading" innovation. It provides a comprehensive description of measurements that exist today, with the intent for the reader to gain a full understanding of the importance of measurements (Badiru and Racz, 2016) and their implications for bolstering innovation in business, industry, government, and the military. Innovation is pervasive and multi-definitional. Let us consider the following:

> Innovation is seeing what everybody has seen and thinking what nobody has thought.
>
> **– Dr. Albert Szent-Gyorgyi**

The quote above is akin to building a new mousetrap after all the conventional mice have been annihilated. Throughout history, humans have striven to come up with better tools, techniques, and instruments for measurement. From the very ancient times to the present fast-paced society, our search for more precise, more convenient, and more accessible measuring devices has led to new developments over the years. The quotes below confirm the importance and efficacy of measurements in our lives. The Appendix presents the most common measurement conversion factors.

> Measure twice, cut once.
>
> **– English Proverb**

> Where there is no Standard there can be no Kaizen (improvement).
>
> **– Taiichi Ohno**

> Where there is no measurement, there can be no standard.
>
> **– Adedeji Badiru**

> Measurement mitigates mess-ups.
>
> **– Adedeji Badiru**

Measurement pervades everything we do. This applies to technical, management, and social activities and requirements. Even in innocuous situations, such as human leisure, the importance of measurement comes to the surface. How much, how far, how good, how fast, and how long are typical conveyances of some sort of measurement. Consider a possible newspaper classified advertisement that provides the following measurement disclaimer.

> The acceptable units of measure for all firewood advertisement are cord or fraction of a cord. The units of measurements for a cord are 4″ by 4″ by 8″. The terms face cord, rack, pile, rick, truckload, or similar terms are not acceptable.

A face cord is a measurement of wood in the United States and Canada. It is a standardized unit of volume that has been defined by law, which means it can be bought or sold with confidence. On the other hand, a cord of wood is one standard length **(4 feet)** of firewood. Who would have thought that firewood had such a serious measurement constraint and guideline? Social, psychological, physical, economic, cognitive, and metabolic attributes, as well as other human characteristics, are all amenable to measurement systems, just as the mechanical devices around us.

Every item that we recognize as innovation is subject to some sort of measurement of its attributes. It could be its length, weight, height, color, volume, time dimension, profit potential, and so on. Consequently, to have confidence in whatever we adjudge to be an innovation, it is essential to think in terms of the pertinent measurements of the characteristics of that item. Therefore, we need to understand what a measurement represents, whether it is on a quantitative scale or a qualitative scale. The measurement must not only be taken but also articulated via communication so that the audience can fully appreciate the item to which it pertains. In this regard, we offer the following inspirational quote:

> While necessity is the mother of invention, communication is the spark for innovation.
>
> **– Adedeji Badiru**

The importance of communication for innovation, from a systems perspective, will be further elucidated in a later section on a systems approach to innovation using the DEJI systems model in Chapter 4.

ANALYSIS OF A MEASUREMENT

It is well understood that we cannot manage anything if we cannot measure it. All elements involved in our day-to-day decision-making require some form of measurement. Measuring an attribute of a system and then analyzing it against some standard, best practice, or benchmark empowers a decision-maker to take appropriate and timely actions.

Fundamentally, measurement is the act, or the result, of a quantitative comparison between a predefined standard and an unknown magnitude. Beckwith and Buck (1965), Shillito and De Marle (1992), Morris (1997), Badiru et al. (2012), and Badiru and Kovach (2012) all address concepts, tools, and techniques of measurement systems. For the result to be generally meaningful, two requirements must be met in the act of measurement:

1. The standard used for comparison must be accurately known and widely accepted.
2. The procedure and instrument employed for obtaining the comparison must be verifiable and repeatable.

The first requirement is that there is an accepted standard of comparison. A weight cannot simply be heavy. It can only be proportionately heavy compared to something else, namely the standard. A comparison must be made, and unless it is made relative to something generally recognized as standard, the measurement can only have limited meaning. This holds for any quantitative measurement we may wish to make. In general, the comparison is one of magnitude, and a numerical result is presupposed. The quantity in question may be twice as large as the standard, or 1.4 times as large, or in some other ratio, but a numerical comparison must be made for it to be meaningful. The typical characteristics of a measurement process include the following:

- Precision
- Accuracy
- Correlation
- Stability
- Linearity
- Type of data

EFFECTS OF WEIGHTS AND MEASURES

> Weights and measures may be ranked among the necessaries of life to every individual of human society. They enter into the economical arrangements and daily concerns of every family. They are necessary to every occupation of human industry; to the distribution and security of every species of property; to every transaction of trade and commerce; to the labors of the husbandman; to the ingenuity of the artificer; to the studies of the philosopher; to the researches of the antiquarian; to the navigation of the mariner, and the marches of the soldier; to all the exchanges of peace, and all the operations of war. The knowledge of them, as in established use, is among the first elements of education and is often learned by those who learn nothing else, not even to read and write. This knowledge is riveted in the memory by the habitual application of it to the employments of men throughout life.
>
> **– John Quincy Adams, Report to the Congress, 1821**

The historical accounts of measurements presented in this section are based mostly on USA. National Institute of Standards and Technology (NIST, 1974), HistoryWorld

(2014), and Ta Neter Foundation (TANF, 2014). Weights and linear measures were among the earliest tools invented by man. Primitive societies needed rudimentary measures for many tasks, such as house and road construction and commerce of raw materials. In early years, humans used parts of the human body (usually the ruling class's own) and the natural surroundings to devise measuring standards. Early Babylonian and Egyptian records and the Bible indicate that length was first measured with the foot, forearm, hand, or finger. Time was measured by the periods of the sun, moon, and other heavenly bodies. When it was necessary to compare the capacities of containers, such as gourds of clay or metal vessels, they were filled with plant seeds that were then counted to measure volumes. With the development of scales as a means for weighing, seeds and stones served as standards. For instance, the "carat," still used as a mass unit for gems, is derived from the carob seed.

As societies evolved, measurements became more complex. The invention of numbering systems and the science of mathematics made it possible to create whole systems of measurement units suitable for trade and commerce, land division, taxation, and scientific research. For these more sophisticated uses, it was necessary not only to weigh and measure more complex items but also to do so accurately and repeatedly at different locations. Because of limited international exchange of goods and communication of ideas in ancient times, different measuring systems evolved for the same measures and became established in different parts of the world. In this same time period, even different parts of the same country might use different measuring systems for the same purpose. Historical records indicate that early measurement systems evolved locally in Africa to take advantage of the African natural environment. For example, common early units of measure in some parts of Africa relied on standardizations based on the size and weight of cocoa beans.

METRICS OF PRODUCTION INNOVATION

In order to assess something, we must be able to measure it. Thus, measurements are important in developing metrics of production innovation. When those measurements are adopted into a model or methodology to ensure quality, gauge outcomes, or meet performance standards in the field of study, they become, *metrics*. This may seem parochial, but it is worth mentioning if this book is to fulfill its purpose. This book is in itself innovative in that it will invite the world to view innovation in a whole new way. No longer will "claims to be innovative" be the marker or the bar. Very soon, and as a result of the metrics being proposed in this book, companies will need to "measure up," by examining, evaluating, and measuring their actions when it comes to innovation, using those results to prove they have innovated, or are innovative. As with nearly everything occurring in our digital era, these measurements will be backed by reliable data analysis. The latter portion of this chapter expands on the considerations of data analysis. Its importance should not be overlooked, and in fact, data analysis should be made paramount when it comes to taking on claiming innovation in the 21st century.

The measurement system commonly used in the United States today is nearly the same as that brought by the American colony settlers from England. These measures had their origins in a variety of cultures, including Babylonian, Egyptian, Roman,

Anglo-Saxon, and Nordic French. The ancient "digit," "palm," "span," and "cubic" units of length slowly lost preference to the length units "inch," "foot," and "yard." Roman contributions include the use of 12 as a base number and the words from which we derive many of the modern names of measurement units. For example, the 12 divisions of the Roman "pes" or foot were called unciae. The "foot," as a unit of measuring length, is divided into 12 inches. The common words "inch" and "ounce" are both derived from the same Latin word, "uncia," meaning one-twelfth. The "yard" as a measure of length can be traced back to early Saxon kings. They wore a sash or girdle around the waist that could be removed and used as a convenient measuring device. Thus, the word "yard" comes from the Saxon word "gird," which represents the circumference of a person's waist, preferably as "standard person," such as a king.

The titles for measurements aside, the evolution and standardization of measurement units often had interesting origins as well. For example, it was recorded that King Henry I decreed that a yard should be the distance from the tip of his nose to the end of his outstretched thumb. The length of a furlong (or furrow-long) was established by early Tudor rulers as 220 of these same King Henry I yards. This led Queen Elizabeth I to declare in the 16th century that the traditional Roman mile of 5,000 feet would be replaced by one of 5,280 feet, making the mile exactly eight furlongs and establishing a convenient relationship between the furlong and the mile. To this day, one mile consists of 5,280 feet or 1,760 yards. Thus, through royal edicts, England had achieved a greater degree of standardization by the 18th century compared to other European countries. The promulgation of English units was improved by the discovery of coal in England and the subsequent transmission of coal to many parts of the world, dawning the First Industrial Revolution. The English units were well suited to burgeoning commerce and trade because they had been developed and refined to meet commercial needs. Through English colonization and its dominance of world commerce during the 17th, 18th, and 19th centuries, the English system of measurement units became established in many parts of the world, including the American colonies. The early 13 American colonies, however, had undesirable differences with respect to measurement standards for commerce. The need for greater uniformity led to clauses in the Articles of Confederation (ratified by the original colonies in 1781) and the Constitution of the United States (ratified in 1788), granting Congress the power to fix uniform standards for weights and measures across the colonies. Today, standards provided by the NIST ensure uniformity of measurement units throughout the country.

The need for a single worldwide coordinated measurement system was recognized over 300 years ago. In 1670, Gabriel Mouton, Vicar of St. Paul's Church in Lyons, France, and an astronomer, proposed a comprehensive decimal measurement system based on the length of one minute of arc of a great circle of the Earth. Mouton also proposed using the swing length of a pendulum with a frequency of one beat per second as the unit of length. A pendulum with this beat would have been fairly easily reproducible, thus facilitating the widespread distribution of uniform standards.

In 1790, in the midst of the French Revolution, the National Assembly of France requested the French Academy of Sciences to "deduce an invariable standard for all the measures and all the weights." The Commission appointed by the Academy created a system that was, all at once, simple and scientific. The unit of length was to be a portion of the Earth's circumference. Measures for capacity (volume) and mass were to be derived from the unit of length, thus relating the basic units of the system to each other and to nature. Furthermore, larger and smaller multiples of each unit were to be created by multiplying or dividing the basic units by 10 and powers of 10. This feature provided a great convenience to users of the system, by eliminating the need for such calculations as dividing by 16 (to convert ounces to pounds) or by 12 (to convert inches to feet). Similar calculations in the metric system could be performed simply by shifting the decimal point. Thus, the metric system is a "base-10" or "decimal" system.

The Commission assigned the name metre (i.e. meter in English) to the unit of length. This name was derived from the Greek word metron, meaning "a measure." The physical standard representing the meter was to be constructed so that it would equal one ten-millionth of the distance from the North Pole to the equator along the meridian running near Dunkirk in France and Barcelona in Spain. The initial metric unit of mass, the "gram," was defined as the mass of one cubic centimeter (a cube that is 0.01 meter on each side) of water at its temperature of maximum density. The cubic decimeter (a cube 0.1 meter on each side) was chosen as the unit for capacity. The fluid volume measurement for the cubic decimeter was given the name "liter." Although the metric system was not accepted with much enthusiasm at first, its adoption by other nations occurred steadily after France made its use compulsory in 1840. The standardized structure and decimal features of the metric system made it well suited for scientific and engineering work. Consequently, it is not surprising that the rapid spread of the system coincided with an age of rapid technological development. In the United States, by the Act of Congress in 1866, it became "lawful throughout the United States of America to employ the weights and measures of the metric system in all contracts, dealings or court proceedings." However, the United States has remained a holdout with respect to a widespread adoption of the metric system. Previous attempts to standardize the metric system in the USA have failed. Even today, in some localities of the USA, both English and Metric systems are used side by side.

As a good illustration of dual use (and subsequent chaos) of measuring systems, a widespread news report in late September 1999 exposed how the National Aeronautics and Space Administration (NASA) lost a $125 million Mars orbiter in a crash onto the surface of Mars because a Lockheed Martin engineering team used the English units of measure while the agency's team used the more conventional metric system for a key operation of the spacecraft. This unit mismatch prevented mutually understood navigation information from being transferred between the Mars Climate Orbiter spacecraft team at Lockheed Martin in Denver and the flight team at NASA's Jet Propulsion Laboratory in Pasadena, California. As this case demonstrates, even high-stakes scientific endeavors are not spared from the havoc that can emerge from not standardizing which units of measurement to use.

Getting back to history, the late 1860s saw the need for even better metric standards to keep pace with scientific advances. In 1875, an international agreement, known as the Meter Convention, set up well-defined metric standards for length and mass and established permanent mechanisms to recommend and adopt further refinements in the metric system. This agreement, commonly called the "Treaty of the

Meter" in the United States, was signed by 17 countries, including the United States. As a result of the Treaty, metric standards were constructed and distributed to each nation that ratified the Convention. Since 1893, the internationally adopted metric standards have served as the fundamental measurement standards of the United States, at least in theory, if not in practice.

By 1900, a total of 35 nations, including the major nations of continental Europe and most of South America, had officially accepted the metric system. In 1960, the General Conference on Weights and Measures, the diplomatic organization made up of the signatory nations to the Meter Convention, adopted an extensive revision and simplification of the system. The following seven units were adopted as the base units for the metric system:

1. Meter (for length)
2. Kilogram (for mass)
3. Second (for time)
4. Ampere (for electric current)
5. Kelvin (for thermodynamic temperature)
6. Mole (for amount of substance)
7. Candela (for luminous intensity)

Based on the general standardization described above, the name Système International d'Unités (International System of Units), with the international abbreviation SI, was adopted for the modern metric system. Throughout the world, measurement science research and development continue to develop more precise and easily reproducible ways of defining measurement units. The working organizations of the General Conference on Weights and Measures coordinate the exchange of information about the use and refinement of the metric system and make recommendations concerning improvements to the system and its related standards. Our daily lives are mostly ruled or governed by the measurements of length, weight, volume, and time, which are briefly described below.

LENGTH MEASUREMENT

The measurement of distance, referred to as length, is the most ubiquitous measurement in our world. The units of length represent how we conduct everyday activities and transactions. The two basic units of length measurement are the British Units (inch, foot, yard, mile) and the metric system (meters, kilometers). In its origin, the inch is anatomically a thumb. The foot logically references the human foot. The yard relates closely to a human pace but also derives from two cubits (the measure of the forearm). The mile originates from the Roman *mille passus*, which means a thousand paces. The ancient Romans defined a pace as two steps. So, approximately, a human takes two paces within one yard. The average human walking speed is about 5 kilometers per hour (km/h), or about 3.1 miles per hour (mph).

For the complex measuring problems of civilization – surveying land to register property rights, or selling a commodity by length – a more precise unit is required. The solution is a rod or bar of an exact length, kept in a central public place. From this 'standard,' other identical rods can be copied and distributed through the community. In Egypt and Mesopotamia, these standards are kept in temples. The basic unit of length in both civilizations is the cubit, based on a forearm measured from elbow to tip of middle finger. When a length such as this is standardized, it is usually the king's dimension which is first taken as the norm.

WEIGHT MEASUREMENT

For measurements of weight, the human body does not provide convenient approximations as it does for length. However, for comparative purposes, grains of wheat turn out to be reasonably standard in size. Weight can be expressed with some degree of accuracy in terms of the number of grains of wheat. The use of grains to convey weight is still used today in the measurement of precious metals, such as gold, by jewelers. Similar to measurements of length, a block of metal can be kept in the temples as the official standard for a given number of grains. Copies of this standard can be cast and weighed on a balance for perfect accuracy. However, due to imperfect human integrity in using scales, it became necessary to establish an inspectorate of weights and measures for the practical adjudication of measurements in ancient times.

WEIGHT AND MASS

It is prudent to remind and distinguish between the measure of "weight" and the measure of "mass." Weight is a unit of force; an item's "weight" on the surface of the earth is determined by the force of gravity being exerted on it. This means that the same item's weight on the moon will differ from its weight on the earth. The earth exerts a downward force on all objects resting on or moving along its surface (or even near it), and it does not matter whether the objects are in motion or at rest, the weight remains the same.

Weight can be calculated by multiplying the mass of an object by gravity, g, or

$$F_g = mg = W$$

The metric units of weight are $kg*m/s^2$.

Mass is the measure of the *inertia* of an object. Inertia is a property of a body or mass such that the object will continue in its current state until it is disrupted by an external force. The units for inertia are $kg*m^2$, where kg is kilogram, the *mass* of the object in metric units, and m^2 is "meter square," or the square of the distance from the axis of rotation to the particles that make up the object. Simply put, the more mass an object has, the greater its inertia and the harder it is to move. To summarize, weight is the unit of *force*, and mass is the unit of *inertia*.

VOLUME MEASUREMENT

From the ancient time of commerce and trading to the present day, a reliable standard of volume has been one of the hardest things to accomplish. However, we improvise by using items from nature and art. Items such as animal skins, baskets, sacks, or pottery jars can be made to approximately consistent sizes, such that they are sufficient for measurements in ancient measurement transactions. When the exact amount of any commodity needs to be known, weight is the measure more likely to be used instead of volume.

TIME MEASUREMENT

Time is a central aspect of human life. Throughout human history, time has been appreciated in very precise terms. Due to celestial precision of day and night, the day and the week are easily recognized and recorded. However, an accurate calendar for the year is more complicated to achieve universally. The morning time before midday (forenoon) is easily distinguishable from the time after midday (afternoon), provided the Sun is shining, and the position of the Sun in the landscape can reveal roughly how much of the day has passed. By contrast, the smaller units of time, such as hours, minutes, and seconds, were initially (in ancient times) unmeasurable and unnecessary. They were unnecessary because ancient human had large blocks of time to complete tasks. Micro-allocation of time was, therefore, not essential. However, in modern society, the minute (tiny) time measurements of seconds and minutes has become essential.

TECHNOLOGY IN MEASUREMENTS

The sundial and water clock originated in the second millennium BC. The movement of the Sun through the sky makes possible a simple estimate of time, from the length and position of a shadow cast by a vertical stick. If marks were made where the Sun's shadow fell, the time of day could be recorded consistently. The result was the sundial. An Egyptian example survives from about 800 BC, though records indicate that astronomers understood the principle much earlier. Practically, it is difficult to measure time precisely on a sundial because the Sun's path through the sky changes with the seasons. Early attempts at precision in timekeeping relied on a different principle known as the water clock. The water clock, derived from the Greek word clepsydra, attempted to measure time by the amount of water which dripped from a tank. This would be a reliable form of clock if the flow of water could be perfectly controlled. In practice, at that time, it could not. The hourglass, using sand on the same principle, had an even longer history and utility. It was a standard feature used on the 18th century pulpits in Britain to ensure a sermon of standard and sufficient duration. The hour as a unit of time measurement originated in the 14th century. Until the arrival of clockwork in the 14th century AD, an hour was a variable concept. It is a practical division of the day into 12 segments (12 being the most convenient number for dividing into fractions, since it is divisible by 2, 3, and 4). For the same reason 60, divisible by 2, 3, 4, and 5, has been a larger framework of measurement ever since the Babylonian times. The traditional concept of the hour, as one-twelfth of the time between dawn and dusk, is useful in terms of everyday timekeeping. Approximate appointments are easily made, at times which are easily sensed. Noon is always the sixth hour. Halfway through the afternoon is the ninth hour. This is famous as the time of the death of Jesus on the Cross. The trouble with the traditional hour is that it differs in length from day to day. And a daytime hour is different from one in the night (also divided into 12 equal hours). A clock cannot reflect this variation, but it can offer something more useful. It can provide every day something which occurs naturally only twice a year, at the spring and autumn equinox, when the 12 hours of day and the 12 hours of night are the same length. In the 14th century, coinciding with the first practical clocks, the meaning of an hour gradually changed. It became a specific amount of time, one twenty-fourth of a full solar cycle from dawn to dawn. Today, the day is recognized as 24 hours, although it still features on clock faces as two twelves.

Minutes and seconds, as we know them today, originated from 14th to 16th centuries. Even the first clocks could measure periods less than an hour, but soon, striking the quarter-hours seemed insufficient. With the arrival of dials for the faces of clocks, in 14th century, something like a minute was required. The middle ages inherited a scale of scientific measurement based on 60 from Babylon. In Medieval Latin, the unit of one sixtieth is *pars minuta prima* ('first very small part'), and a sixtieth of that is *pars minute secunda* ('second very small part'). Thus, on a principle that is 3,000 years old, minutes and seconds have made their way into our modern time. Minutes were mentioned from the 14th century, but clocks were not precise enough for "seconds" of time to be needed until two centuries later.

Hero's Dioptra was from the first century AD. One of the surviving books of Hero of Alexandria, entitled *On the Dioptra*, describes a sophisticated technique which he developed for surveying land. Plotting the relative position of features in a landscape (essential for any accurate map) is a more complex task than simply measuring distances. It was necessary to discover accurate angles in both the horizontal and vertical planes. To make this possible, a surveying instrument had to consistently maintain both planes in different places, to take readings of the deviation in each plane between one location and another. This is what Hero achieved with the instrument mentioned in his title, the *dioptra*, which approximately means, the 'spyhole,' through which the surveyor looks when pinpointing the target in order to read the angles. For his device, Hero adapted an instrument long used by Greek astronomers (e.g., Hipparchus) for measuring the angle of stars in the sky. In his day, Hero achieved his device without the convenience of two modern inventions, the compass and the telescope.

The barometer and atmospheric pressure originated during 1643–1646. Like many significant discoveries, the principle of the barometer is observed by accident. Evangelista Torricelli, assistant to Galileo at the end of his life, was curious about why it is more difficult to pump water from a well in which the water lies far below ground level. He suspected that the reason might be the weight of the extra column of air above the water, and he devised a way to test this theory. He filled a glass tube with mercury, an element that is 13.6 times denser than water, thus saving Torricelli from needing a tube greater than 30 feet high. Submerging it in a bath of mercury and raising the sealed end to a vertical position, he found that the mercury dropped slightly down the tube. He reasoned that the weight of air on mercury in the bath was supporting the weight of the column of mercury in the tube. If this was true, then the space in the glass tube above the mercury column must be a vacuum. This led him into controversy with traditional scientists of the day, who believed nature abhorred a vacuum. However, it also encouraged von Guericke, a decade later, to develop the vacuum pump. The concept of varying atmospheric pressure occurred to Torricelli in 1643, when he noticed that the height of his mercury column sometimes varied slightly from its normal level, which was 760 mm above the mercury level in the bath. His observation suggested that these variations closely corelated with changes in the weather. This was the origin of the barometer. With the concept thus established that air had weight, Torricelli was able to predict there must be less atmospheric pressure at higher altitudes. In 1646, Blaise Pascal, aided by his brother-in-law, carried a barometer to different levels of the 4,000 foot mountain Puy de Dôme, near Clermont, to take readings. The confirmation was that atmospheric pressure varied with altitude.

The mercury thermometer originated circa 1714–1742. Gabriel Daniel Fahrenheit, a German glassblower and instrument maker working in Holland, was interested in improving the design of the thermometer that had been in use for half a century. Known as the Florentine thermometer (because it was developed in the 1650s in Florence's Accademia del Cimento), this pioneering instrument depended on the expansion and contraction of alcohol within a glass tube. Alcohol expands rapidly with an increase in temperature, but not at an entirely regular speed of expansion. This made accurate readings difficult, as also did the sheer technical problem of blowing glass tubes with very narrow and entirely consistent bores. By 1714, Fahrenheit had made great progress on the technical front, creating two separate alcohol thermometers which agreed precisely in their reading of temperature. In that year, he heard of the research of a French physicist, Guillaume Amontons, into the thermal properties of mercury. Mercury expands less than alcohol (about seven times less for the same increase in temperature), but it does so in a more regular manner. Fahrenheit saw the advantage of this regularity, and he used his glass-making skills to fashion a glass bore that accommodated the smaller rate of expansion. He constructed the first mercury thermometer that subsequently became a standard. There remained the problem of how to calibrate the thermometer to show degrees of temperature. The only practical method was to choose two temperatures which could be established independently and mark them on the thermometer and divide the intervening length of the tube into several equal degrees. In 1701, Sir Isaac Newton proposed the freezing point of water for the bottom of the scale and the temperature of the human body for the top end. Fahrenheit, accustomed to Holland's cold winters, wanted to include temperatures below the freezing point of water. He, therefore, used blood temperature for the top of his scale but adopted the freezing point of salt water for the lower extreme.

Measurement is conventionally done in multiples of 2, 3, and 4, so Fahrenheit split his scale into 12 sections, each of them divided into eight equal parts. This gave him a total of 96°, zero being the freezing point of brine and 96° as an inaccurate estimate of the average temperature of the human blood. The actual mean human body temperature is 98.6°. With his thermometer calibrated on these two points, Fahrenheit could take a reading for the freezing point (32°) and boiling point (212°) of water. In 1742, a Swede, Anders Celsius, proposed an early example of decimalization. His centigrade scale took the freezing and boiling temperatures of water as 0° and 100°. In English-speaking countries, this less- complex system took more than two centuries to be embraced. Even today, the Fahrenheit unit of temperature is more prevalent in some countries, such as the United States.

The chronometer was developed circa 1714–1766. Two centuries of ocean travel had made it essential for ships on naval or merchant business to calculate their positions accurately in any of the oceans in the world. With the help of the simple and ancient astrolabe, the stars would reveal latitude. However, on a revolving planet, determining longitude was more challenging. It was essential to know what time it was before it could be determined what place it was. The importance of this was made evident in 1714 when the British government established a Board of Longitude and offered a massive prize of £20,000 to any inventor who could produce a clock capable of keeping accurate time at sea. The terms were demanding. To win the prize, a chronometer needed to be sufficiently accurate to calculate longitude within 30 nautical miles at the end of a journey to the West Indies. This meant that in rough seas, damp salty conditions, and sudden changes in temperature, the instrument must lose or gain not more than three seconds per day. This was a level of accuracy unmatched at the time by the best clocks. The challenge appealed to John Harrison, a 21-year-old Lincolnshire carpenter with an interest in clocks. It was nearly 60 years passed before he won the money. Fortunately, he lived long enough to collect it. By 1735, Harrison had built the first chronometer, which he believed approached the necessary standard. Over the next quarter-century, he replaced it with three improved models before formally undergoing the government's test. His innovations included bearings which reduced friction, weighted balances interconnected by coiled springs to minimize the effects of movement, and the use of two metals in the balance spring to cope with expansion and contraction caused by changes in temperature. Harrison's first 'sea clock', built in 1735, weighed 72 pounds and was 3 feet in all dimensions. His fourth, in 1759, was more like a watch, being circular and 5 inches in diameter. This version underwent the sea trials. Harrison was 67 years old, so, his son took the chronometer on its test journey to Jamaica in 1761. It was 5 seconds slow at the end of the voyage. The government argued that this result might have been a fluke and offered Harrison only £2,500. After further trials, and the successful building of Harrison chronometer by another craftsman (at the huge cost of £450), the inventor was finally paid the full prize money in 1773.

Harrison demonstrated in 1761 what was possible, but his chronometer was an elaborate and expensive way to achieve the goal. It was in France, where a large prize was also offered by the Académie des Sciences, where the practical chronometer of the future was developed. The French trial, open to all comers, took place in 1766 on a voyage from Le Havre in a specially commissioned yacht, the *Aurore*. The only chronometer ready for the test was designed by Pierre Le Roy. At the end of 46 days, his machine was accurate to within 8 seconds. Le Roy's timepiece was larger than Harrison's final model, but it was much easier to construct. It provided the pattern of the future. With further modifications from various sources over the next two decades, the marine chronometer emerged before the end of the 18th century. Used in combination with the sextant, it explorers traveling the world's oceans to bring back accurate information of immense value to the makers of maps and charts of the world.

The sextant originated between 1731 and 1757. The 18th century search for a way to discover longitude was accompanied by refinements in the ancient method of establishing latitude. This had been possible since the 2nd century BC by means of the *astrolabe*. From the beginning of European voyages in the 15th century, practical improvements were made to the astrolabe, mainly by providing more convenient calibrated arcs on which the user could read the number of degrees of the sun or a star above the horizon. The size of these arcs was defined in relation to the full circle. A quadrant (a quarter of the circle) showed 90°, a sextant 60°, and an octant 45°. The use of such arcs in conjunction with the traditional astrolabe is evident from a text dating back to 1555 that related about voyages to the West Indies. The author referred to "quadrant and astrolabe, instruments of astronomy." An important "innovative" improvement during the 18th century was the application of optical devices (mirrors and lenses) to promote the task of working out angles above the horizon. Slightly different solutions emerged from instrument makers in Europe and America, which competed during the early decades of the century. The one which prevailed, mainly because it was more convenient at sea, was designed as an octant in 1731 by John Hadley, an established English maker of reflecting telescopes. Hadley's instrument, like others designed by his contemporaries, used mirrors to bring any two points into alignment in the observer's sight-line. For the navigator at that time, these two points would usually be the sun and the horizon. To read the angle of the sun, the observer looked through the octant's eyepiece at the horizon and then turned an adjusting knob until the reflected sphere of the sun (through a darkened glass) was brought down to the same level. The double reflection meant that the actual angle of the sun above the horizon was twice that of the octant's arc of 45%. So, Hadley's instrument could read angles up to 90%. In 1734, Hadley added an improvement which became the standard by installing an additional level so that the horizontal could be found even if the horizon was not visible. In 1757, after Hadley's death, a naval captain proposed that the arc in the instrument be extended from 45° to 60°, making possible a reading up to 120°. With this, Hadley's octant became a sextant, and the instrument came into general use since then.

Africa is home to the world's earliest known use of measurement and calculation, confirming the continent as the origin of both basic and advanced mathematics. Thousands of years ago, while parallel developments were going on in Europe, Africans were using rudimentary numerals, algebra, and geometry in daily life. This knowledge spread throughout the entire world after a series of migrations out of Africa, beginning around 30,000 BC, and later following a series of invasions of Africa by Europeans and Asians (1700 BC-present). It is historically documented that early man migrated out of Africa to Europe and Asia. This feat of early travel and navigation may have been facilitated by the indigenous measurement systems of ancient Africa. The following sections recount measuring and counting in ancient Africa.

PREHISTORIC TECHNOLOGY

The oldest known mathematical instrument is the Lebombo bone, a baboon fibula used as a measuring device and so named for its location of discovery in the Lebombo mountains of Swaziland. The device is at least 35,000 years old. Judging from its 29 distinct markings, it may have been used to either track lunar cycles or as a measuring stick. It is rather interesting to note the significance of the 29 markings (roughly the same number as the lunar cycle, i.e., 29.531 days) on the baboon fibula because it is the oldest indication that the baboon, a primate indigenous to Africa, was symbolically linked to Khonsu, who was also associated with time. The Kemetic god, Djehuty ("Tehuti" or "Toth"), was later depicted as a baboon or an ibis, which is a large tropical wading bird with a long neck and long legs. This animal symbolism is usually associated with the moon, math, writing, and science. The use of baboon bones as measuring devices was continuous across Africa, suggesting that Africans always held the baboon as sacred and associated with the moon, math, and time.

The world's oldest evidence of advanced mathematics is also a baboon fibula, discovered in present-day Democratic Republic of Congo, and dates to at least 20,000 BC. The bone is now housed in the Museum of Natural Sciences in Brussels. The Ishango bone is not merely a measuring device or tally stick as some people erroneously suggest. The bone's inscriptions are clearly separated into clusters of markings that represent various quantities. When the markings are counted, they are all odd numbers with the left column containing all prime numbers between 10 and 20, and the right column containing added and subtracted numbers. When both columns are calculated, they add up to 60 (nearly double the length of the lunar cycle). We recall that the number 60 also featured prominently in the development of early measuring devices in Europe.

Although the oldest known evidence of the ancient counting board game, Gebet'a (or "Mancala" as it is more popularly known) comes from Yeha (700 BC) in Ethiopia, it was probably used in Central Africa many years prior to that. The game forced players to strategically capture a greater number of stones than one's opponent. The game usually consists of a wooden board with two rows of six holes each, and two larger holes at either end. However, in antiquity, the holes were more likely to be carved into stone, clay, or mud. More advanced versions found in Central and East Africa, such as the Omweso, Igisoro, and Bao, usually involve four rows of eight holes each.

Housed in Moscow's Pushkin State Museum of Fine Arts, the so-called "Moscow" papyrus, was purchased by Vladimir Golenishchev sometime in the 1890s. Written in hieratics from, perhaps, the 13th dynasty in Kemet, the name of ancient Egypt, the papyrus is one of the world's oldest examples of the use of geometry and algebra. The document contains approximately 25 mathematical problems, including the calculation for the length of a ship's rudder, the surface area of a basket, the volume of a frustum (a truncated pyramid), and various ways of solving for unknowns. One feature of the papyrus is a number with a bar over it, such as, which represents the unit fraction of 1/3. Fractions were important to the Egyptians because it allowed them to divide objects for construction (such as bricks for the pyramids) to coincide with the number of workers. There were tens of thousands of workers, so this task was mathematically essential.

Purchased by Alexander Rhind in 1858 AD, the so-called "Rhind" Mathematical papyrus dates to approximately 1650 BC and is presently housed in the British Museum. Although some Egyptologists link this to the foreign Hyksos, this text was found during excavations at the Ramesseum in Waset (Thebes) in Southern Egypt, which never came under Hyksos' rule. The first page contains 20 arithmetic problems, including the addition and multiplication of fractions, and 20 algebraic problems, including linear equations. The second page shows how to calculate the volume of rectangular and cylindrical granaries, with pi (Π) estimated at 3.1605. There are also calculations for the area of triangles (slopes of a pyramid) and an octagon. The third page continues with 24 problems, including the multiplication of algebraic fractions, among others. Notably, these two papyri are separated by approximately 200 years, and the Rhind papyrus is considerably more detailed. Using today's pace of discovery and invention, this same additional detail would have been achieved in 0.3 seconds. The accelerated pace of discovery is a key component of innovation, which will be explained in more detail in a subsequent chapter.

Timbuktu in Mali is well-known as a hub of commerce in ancient times. Timbuktu is home to one of the world's oldest universities, Sankore, which had libraries full of manuscripts mainly written in Ajami (African languages, such as Ajami and Hausa, were written in a script similar to "Arabic") in the 1200s AD. When Europeans and Western Asians began visiting and colonizing Mali from 1300s to 1800s AD, Malians began to hide the manuscripts, as they were seen as heresy by certain religious groups visiting the region. Many of the scripts were mathematical and astronomical. In recent years, as many as 700,000 scripts have been rediscovered and attest to the continuous knowledge of advanced mathematics, science, and measurements in Africa, well before European colonization.

TECHNOLOGY ASSESSMENT

There are two basic methods of measurement:

1. *direct comparison* with either a primary or a secondary standard
2. *indirect comparison* with a standard through a calibrated system.

Direct comparison: how do you measure the length of a cold-rolled bar? You probably use a steel tape. You compare the bar's length with a standard. The bar is so many feet long because that many units on your standard has the same length as the bar. You have determined this by making a direct comparison. Although you cannot access to the primary standard defining the unit, you can effectively rely on a secondary standard. Primary measurement standards have the least amount of uncertainty compared to the certified value and are traceable directly to the SI. Secondary standards, on the other hand, are derived by ratifying their accuracy through direct comparison to a primary standard.

In some respect, measurement by direct comparison is quite common. Many length measurements are made in this way, using a ruler or tape measure, a tool considered to be a primary standard. In addition, the time of day is usually determined by glancing at one's watch (indirect comparison) and using the watch used as a secondary standard. The watch goes through its double-cycle (two 12 hour periods) in synchronization with the Earth's rotation. Although, in this case, the primary standard is available to everyone (looking up in the sky at the sun from the earth), the watch is more convenient because it works on cloudy days, indoors, outdoors, in daylight, and in the dark (at night). It is also more precise. That is, its resolution is better. In addition, if well-regulated, the watch is more accurate, because the earth does not rotate at a uniform speed. It is seen, therefore, that in some cases, a secondary standard is more useful than the primary standard.

Measuring by direct comparison implies stripping the measurement problem to its barest essentials. However, the method is not always the most accurate or the best. The human senses are not uniformly capable to make direct comparisons for all quantities with equal facility, and in many cases, they are not sensitive enough. We can make direct length comparisons using a steel rule with a level of precision of about 0.01 inch. However, greater accuracy is required, in which case we must call for additional assistance from some calibrated measuring system.

Indirect comparison: while we can do a reasonable job through direct comparison of length, how well can we compare masses, for example? Our senses enable us to make rough comparisons. We can lift a pound of meat and compare its effect with an unknown mass. If the unknown mass is similar, we may be able to say that it is slightly heavier or lighter than our "standard" pound. However, we could never be certain that the two masses were the same, even within one ounce. Our ability to make this comparison is not as good as it is for the displacement of the mass. Our effectiveness in coming close to the standard is related to our ability to "gauge" the relative impacts of mass on our ability to displace the mass. This brings to mind the common riddles of "Which weighs more? A pound of feathers or a pound of stones?" Of course, both weigh the same with respect to the standard weight of "pound."

In making most engineering measurements, we require the assistance of some form of measuring system, and measurement by direct comparison is less general than measurement by indirect comparison. Most mechanical measurement systems (Beckwith and Buck, 1965) fall within the framework of a generalized arrangement consisting of three stages, as follows:

Stage I:	A detector-transducer stage
Stage II:	An intermediate modifying stage
Stage III:	The terminating stage, consisting of one or a combination of an indicator, a recorder, or some form of controller

Each stage is consists of a distinct component or grouping of components which perform required and definite steps in the measurement. These components may be termed basic elements, whose scope is determined by their function rather than their construction. First stage detector-transducer: The primary function of the first stage is to detect or sense the input signal. This primary device must be sensitive to the input quantity. At the same time, ideally it should be insensitive to every other possible input. For instance, if it is a pressure pickup, it should not be sensitive to, acceleration; if it is a strain gauge, it should be insensitive to temperature; or if a linear accelerometer, it should be insensitive to angular acceleration, and so on. Unfortunately, it is extremely rare to find a detecting device that is completely selective. As an example of a simple detector-transducer device, consider an automobile tire pressure gauge. It consists of a cylinder and a piston, a spring resisting the piston movement, and a stem with scale divisions. As the air pressure bears against the piston, the resulting force

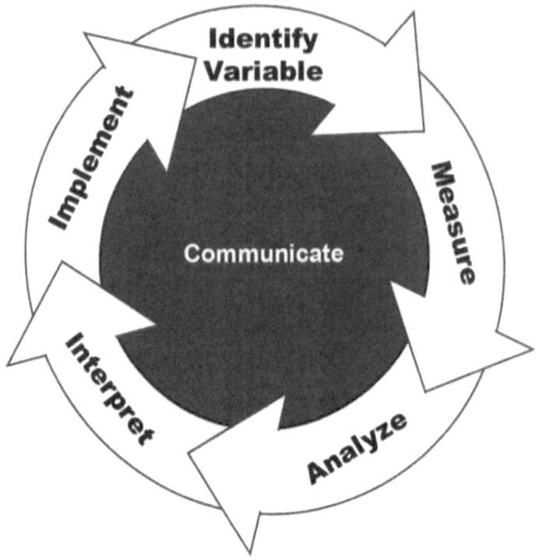

FIGURE 7.1 Generic Measurement Loop.

compresses the spring until the spring and air forces are balanced. The calibrated stem, which remains in place after the spring returns the piston, indicates the applied pressure. In this case, the piston–cylinder combination along with the spring forms the detector-transducer. The piston and cylinder form one basic element, while the spring is another basic element. The piston–cylinder combination, serving as a force-summing device, senses the pressure effect, and the spring transduces it into the displacement. Realistically, not all measurements we encounter in theory and practice are of transduceable mechanical settings. Therefore, measurements can take more generic paths of actualization. Figure 7.1 shows a generic measurement loop revolving around variable identification, actual measurement, analyzing the measurement result, interpreting the measuring in the context of the prevailing practical application, and implementing the measurement for actionable decisions. In each stage of the loop, communication is a central requirement. Communication can be in the form of a pictorial display, a verbal announcement, or a written dissemination. A measurement is not usable unless it is communicated in an appropriate format and at the appropriate time.

Every decision requires data collection, measurement, and analysis. In practice, we encounter different types of measurement scales depending on the particular items of interest. Data may need to be collected on decision factors, costs, performance levels, outputs, and so on. The different types of data measurement scales that are applicable are presented below.

NOMINAL SCALE OF MEASUREMENT

The nominal scale is the lowest level of measurement scales. It classifies items into categories, that are mutually exclusive and collectively exhaustive. This means, that the categories do not overlap and they cover all possible categories of the characteristics being observed. For example, in the analysis of the critical path in a project network, each job is classified as either critical or not critical. Gender, type of industry, job classification, and color are examples of measurements on a nominal scale.

ORDINAL SCALE OF MEASUREMENT

The ordinal scale is distinguished from a nominal scale by the property of order among the categories. For example, in prioritizing project tasks for resource allocation, we know that first is above second, but we do not know how far above. Similarly, we know that better is preferred to good, but we do not know by how much. In quality control, the ABC classification of items based on the Pareto distribution is an example of a measurement on an ordinal scale.

INTERVAL SCALE OF MEASUREMENT

The interval scale is distinguished from the ordinal scale by having equal intervals between the units of measurement. The assignment of priority ratings to project objectives on a scale of 0 to 10 is an instance of a measurement on an interval scale. Even though an objective may have a priority rating of zero, it does not mean that the objective holds no significance to the project team. Similarly, a zero score on an examination does not indicate that a student has no knowledge of materials covered by the examination. Temperature is a key example of an item that is measured on an interval scale. Although there is a zero point on the temperature scale, it is an arbitrary relative measure. Other examples of interval scale include IQ measurements and aptitude ratings.

RATIO SCALE MEASUREMENT

The ratio scale has the same properties as the interval scale, but with a true zero point. For example, an estimate of zero time unit for the duration of a task is a ratio scale measurement. Other examples of items measured on a ratio scale are cost, time, volume, length, height, weight, and inventory level. Many of the items measured in engineering systems will be on a ratio scale.

Another important aspect of measurement involves the classification scheme used. Most systems contain both quantitative and qualitative data. Quantitative data require describing the characteristics of the items being studied numerically. Qualitative data, on the other hand, are associated with attributes that are not measured numerically. Most items measured on the nominal and ordinal scales are normally be classified into the qualitative data category while those measured on the interval and ratio scales are normally be classified into the quantitative data category. The implication for engineering system control is that qualitative data can lead to bias in the control mechanism because qualitative data are subject to the personal views and interpretations of the person using the data. As much as possible, data for an engineering systems control should be based on a quantitative measurement. Figure 7.2 illustrates the four different types of data classification. Notice that temperature is included in the "relative" category rather the "true zero" category. Although zero temperature points on the common temperature scales (i.e., Fahrenheit,

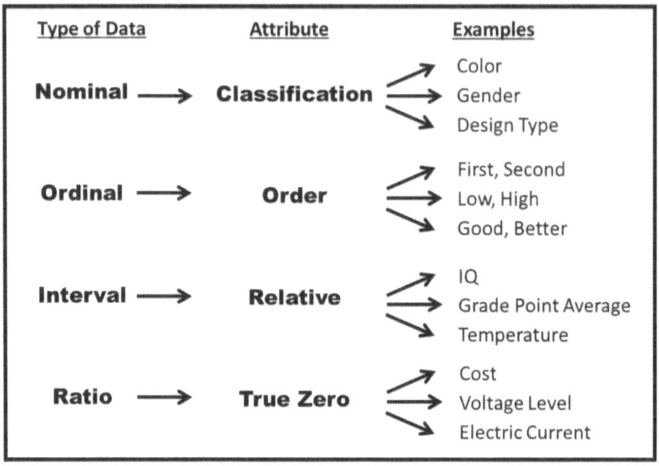

FIGURE 7.2 Four Primary Types of Data Measurement Scales.

Celsius, and Kelvin), these points are experimentally or theoretically established. They are not true zero points, as one might find in a counting system.

Apart from the total and average of a data set, other data properties of interest for innovation metrics and assessment include mode, median, range, variance, deviation, quartiles, and percentiles of the data set. The median is a position measure because its value is determined by its position within the set of observations. Other measures of position include *quartiles* and *percentiles*. There are three quartiles, which divide a set of data into four equal categories. The first quartile, denoted Q_1, is the value below which one-fourth of all the observations in the data set fall. The second quartile, denoted, Q_2, is the value below which two-fourths or one-half of all the observations in the data set fall. The third quartile, denoted Q_3, is the value below which three-fourths of the observations fall. The second quartile is identical to the median. It is technically incorrect to refer to a fourth quartile because this would imply that there is a point within the data set below which all the data points fall: a contradiction! A data point cannot lie within the range of the observations and at the same time exceed all the observations, including itself.

The concept of percentiles is similar to the concept of quartiles except that reference is made to percentage points. There are 99 percentiles that divide a set of observations into 100 equal parts. The *X* percentile is the value below which *X* percent of the data fall. The 99 percentile refers to the point below which 99 percent of the observations fall. The three quartiles discussed previously are regarded as the 25th, 50th, and 75th percentiles. It would be technically incorrect to refer to the 100th percentile. In performance ratings, such as on an examination or product quality level, the higher the percentile of an individual or product, the better. In many cases, recorded data are classified into categories that are not indexed to numerical measures. In such cases, other measures of central tendency or position will be needed. The mode is defined as the value that has the highest frequency in a set of observations. When the recorded observations can only be classified into categories, the mode can be particularly helpful in describing the data. Given a set of K observations (e.g., revenues), X_1, X_2, X_3, ..., X_K, the mode is identified as that value that occurs more than any other value in the set. Sometimes, the mode is not unique in a set of observations.

The results of data analysis can be reviewed directly to determine where and when project control actions may be needed. The results can also be used to generate control charts.

HUMAN INTERFACE IN MEASUREMENTS

Human interfaces with measurement systems and tools date back to ancient times. Even today, we must recognize past approaches to measurements, appreciate present practices of measurements, and be prepared for future developments in measurements for the purpose of driving and advancing innovation. To illustrate this point home, the Angbuilgu is a sundial that was used during the Josean Dynasty in Korea. The name "Angbuilgu" means "upward-looking kettle that catches the shadow of the sun." The 13 horizontal lines mark the 24 periods of seasonal change from the winter solstice to the summer solstice and allow the season of the year to be determined. The vertical lines indicate time, or where one is in time. The device is aligned to face the North Star, and it is marked with pictures of animals (rather than letters), making it accessible to the common people who could not read. The sundial tool was first created in 1434, but the later Angbuilgu was created in the latter half of the 17th century. This distinctive, historical tool that advanced the telling of time beyond anything that had come before it demonstrates that innovation by humans has been changing the world for generations. Now is the time to think of developing, communicating, understanding, and utilizing measurements for innovation.

REFERENCES

Badiru, A. B., Oye Ibidapo-Obe, and B. J. Ayeni (2012), *Industrial Control Systems: Mathematical and Statistical Models and Techniques*, Taylor & Francis Group/CRC Press, Boca Raton, FL.

Badiru, A. B. and Tina Kovach (2012), *Statistical Techniques for Project Control*, Taylor & Francis Group/CRC Press, Boca Raton, FL.

Badiru, A. B. and LeeAnn Racz (2016), *Handbook of Measurements: Benchmarks for Systems Accuracy and Precision*, Taylor & Francis Group/CRC Press, Boca Raton, FL.

Beckwith, T. G. and Buck, N. L. (1965), *Mechanical Measurements*, Addison-Wesley Publishing Company, Reading, MA.

HistoryWorld (2014), http://www.historyworld.net/wrldhis/PlainTextHistories.asp?historyid=ac07, accessed December 24, 2014.

Morris, A. S. (1997), *Measurement and Calibration Requirements for Quality Assurance to ISO 9000*, John Wiley & Sons, Inc., New York, NY.

NIST (1974), NBS Special Publication; 304A (Stock Number 003-003-03501-7), September 1974 U.S. DEPARTMENT OF COMMERCE Technical Administration National Institute Of Standards and Technology, Washington, DC.

Shillito, M. L. and De Marle, D. J. (1992), *Value: Its Measurement, Design, and Management*, John Wiley & Sons, Inc., New York, NY.

TANF (2014), "Restoring Africa's Lost Legacy," http://www.taneter.org/math.html, TA Neter Foundation, accessed December 23, 2014.

8 AI Technology and Predictive Maintenance*

By Thomas L. Wiese

* Reprinted under open access CC BY license from: Wiese, Thomas L. (2024), "Predictive Maintenance Using Artificial Intelligence in Critical Infrastructure: A Decision-Making Framework," *International Journal of Engineering, Business and Management (IJEBM)*, Vol. 8, Issue 4, Oct-Dec, 2024, pp. 1-4. DOI: https://doi.org/10.22161/ijebm.8.4.1.

INTRODUCTION

Critical infrastructure systems are vital for maintaining societal functions, from electricity distribution and transportation networks to water management and telecommunications. The failure of assets in these systems can lead to significant disruptions, jeopardizing safety and incurring substantial financial costs (Olejnik et al., 2020). Maintenance strategies have evolved from reactive and preventive approaches to more predictive and proactive ones, where data analytics and AI can play crucial roles (Zonta et al., 2020). Predictive maintenance (PdM) powered by AI offers an advanced method of optimizing asset lifespan, reducing downtime, and minimizing maintenance costs (Nguyen et al., 2022).

This chapter aims to provide a comprehensive review of AI-driven PdM in critical infrastructure, presenting a decision-making framework that assists organizations in determining when AI-based PdM should be implemented. This framework covers technical, economic, and regulatory aspects of AI adoption.

BACKGROUND AND CONTEXT

Predictive Maintenance

Predictive maintenance refers to the practice of using real-time data and historical trends to predict when equipment is likely to fail, allowing for timely intervention to avoid failures (Carvalho et al., 2019). Unlike preventive maintenance, which relies on fixed schedules, PdM optimizes maintenance efforts based on actual asset conditions, thus reducing unnecessary interventions (Jardine et al., 2006). PdM is especially relevant in critical infrastructure, where failures can have catastrophic effects on safety and service delivery (Ahmad & Kamaruddin, 2012).

 DOI: 10.1201/9781032659220-8

AI and Machine Learning in Maintenance

AI, particularly machine learning (ML) and deep learning (DL), is transforming PdM by improving the accuracy of failure predictions and providing more precise insights into asset health (Zhang et al., 2019). AI-based systems can process vast amounts of sensor data to identify subtle patterns that might indicate impending equipment failures, which traditional rule-based systems may miss (Schwabacher & Goebel, 2007).

ML models, such as random forests and support vector machines, are widely used for PdM due to their ability to handle high-dimensional sensor data (Kusiak et al., 2013). DL models, including neural networks, can capture more complex failure modes, making them effective for handling large datasets from critical infrastructure sectors like energy and water management (Ren et al., 2021).

Critical Infrastructure Sectors

Energy Sector: AI-based PdM is increasingly being adopted in the energy sector, where failures can disrupt power supply and create ripple effects across industries. AI models are used to predict failures in wind turbines, power transformers, and grid components (Zhou et al., 2022). For instance, research shows that AI-based PdM can extend the lifespan of wind turbines by 20%, reducing maintenance costs by up to 30% (Heidari et al., 2021).

Transportation: Railways, airports, and road networks are deploying AI to monitor equipment health. AI models analyze vibration, temperature, and acoustic data to predict issues with trains, tracks, and airplanes, improving safety and reducing maintenance delays (Zio, 2013). A study on the European railway system found that AI-driven PdM increased reliability by 15–20% (Fumeo et al., 2015).

Water Management: AI-based PdM in water infrastructure can prevent costly failures in pumps, pipelines, and treatment plants. Predicting leaks or pump malfunctions allows water utilities to avoid unplanned shutdowns while ensuring continuous service and compliance with water quality standards (Van Thienen et al., 2020). Research has shown that PdM can reduce water utility operational costs by 25% (Na et al., 2020).

Telecommunications: AI systems have become indispensable in telecommunications, where network reliability is paramount. PdM in this sector uses AI to predict failures in data transmission hardware, ensuring uninterrupted service and improving customer satisfaction (Zhao et al., 2020).

CRITICAL INFRASTRUCTURE

Energy Sector

AI-based PdM has seen wide adoption in the energy sector, particularly in renewable energy production. Wind turbines and solar panels, for example, generate massive streams of data that can be analyzed to predict faults (Zhou et al., 2022). AI models, including DL and recurrent neural networks, are used to analyze real-time sensor data for early detection of failures in turbine components, such as bearings and blades (Ren et al., 2021). This has led to a significant reduction in maintenance costs and an increase in the operational efficiency of energy assets (Nguyen et al., 2022).

Transportation

In the transportation sector, AI-driven PdM is used to ensure the safety and reliability of trains, airplanes, and other vehicles. For example, AI systems monitor the condition of railway tracks and train components using data from vibration and acoustic sensors (Zio, 2013). Studies in the European rail sector demonstrate that AI-based PdM can reduce unplanned maintenance by up to 20% (Fumeo et al., 2015).

Water Management

In water management systems, AI-based PdM helps predict equipment failures in pumps, pipelines, and water treatment facilities (Van Thienen et al., 2020). Sensor data, including pressure and flow rate readings, are analyzed by AI models to detect anomalies that could indicate imminent failures, allowing for preemptive action (Na et al., 2020). This reduces downtime and prevents disruptions in water service.

Telecommunications

In telecommunications networks, AI-based PdM is used to predict failures in both hardware and software systems. By analyzing data from network components, AI can detect patterns that may indicate a risk of failure, allowing for preventive actions to ensure network uptime (Zhao et al., 2020). Studies have found that AI-driven PdM can reduce downtime by 15% in telecommunications networks (Ren et al., 2021).

DECISION-MAKING FRAMEWORK

Given the benefits of AI-driven PdM, it is essential for organizations to have a structured approach to evaluate when and how AI should be implemented. The proposed framework consists of four major stages: technical feasibility, economic analysis, regulatory and safety considerations, and pilot testing and scalability. The following framework is proposed in order to determine the applicability of using AI applications in critical infrastructure.

Technical Feasibility Assessment

The first step is to assess whether the necessary technical infrastructure is in place. This includes ensuring that the organization has:

- **Data Availability**: Sufficient data must be available from operational sensors (Zhang et al., 2019).
- **Data Quality**: The quality of the data is critical, as poor data can lead to unreliable AI predictions (Heidari et al., 2021).

- **Computational Resources**: The organization needs adequate computational power to process the data in real time.

(Zhou et al., 2022)

Economic Analysis

An economic analysis is crucial to justify the adoption of AI-based PdM. This involves:

- **Upfront Costs**: The costs of implementing AI, including hardware, software, and personnel training, must be considered (Na et al., 2020).
- **Operational Savings**: AI-based PdM should provide long-term savings by reducing downtime and optimizing maintenance schedules (Carvalho et al., 2019).
- **ROI Calculation**: A thorough cost-benefit analysis should calculate the return on investment over the asset lifecycle.

(Schwabacher & Goebel, 2007)

Regulatory and Safety Considerations

AI-based PdM must comply with the relevant industry regulations and safety standards. This includes ensuring:

- **Compliance with Regulatory Standards**: AI systems should meet legal requirements for data security and operational safety (Zhao et al., 2020).
- **Explainability of AI Models**: AI systems should be transparent, allowing operators to understand and trust their predictions.

(Zhou et al., 2022)

Pilot Testing and Scalability

Pilot testing allows for the evaluation of AI-based PdM in a limited scope before full-scale deployment:

- **Pilot Program Design**: Select a subset of assets or geographical area for pilot testing (Kusiak et al., 2013).
- **Scalability**: Ensure that the system can scale across the entire network without performance degradation.

(Van Thienen et al., 2020)

CHALLENGES AND FUTURE DIRECTIONS

While AI-based PdM holds promise, several challenges remain. Data quality and availability are paramount to success, as poor data can undermine the accuracy of predictions (Ren et al., 2021). Additionally, the complexity of AI models makes it difficult for operators to interpret the results, leading to a reluctance to adopt these systems fully (Zhang et al., 2019). Future research will focus on improving explainability, integrating reinforcement learning, and creating hybrid models that combine ML with Physics-based approaches (Heidari et al., 2021).

CONCLUSION

AI-based PdM is transforming maintenance strategies in critical infrastructure by providing accurate failure predictions and optimizing maintenance schedules. However, its adoption requires careful evaluation through a structured decision-making framework. This framework, covering technical feasibility, economic viability, regulatory considerations, and scalability, provides decision-makers with a systematic approach to deploying AI for PdM. As AI technology advances, its role in improving the reliability and efficiency of critical infrastructure will continue to expand.

REFERENCES

Ahmad, R., & Kamaruddin, S. (2012). An overview of time- based and condition-based maintenance in industrial application. *Computers & Industrial Engineering*, 63(1), 135–149.
Carvalho, T. P., et al. (2019). A review of machine learning and Internet of Things (IoT) in smart maintenance. *Computers in Industry*, 107, 100–117.
Fumeo, E., et al. (2015). Machine learning applications in railway asset management: A review. *Transportation Research Part C: Emerging Technologies*, 72, 53–67.
Heidari, M., et al. (2021). AI-powered predictive maintenance in renewable energy systems: A review. *Renewable and Sustainable Energy Reviews*, 135, 110239.
Jardine, A. K., Lin, D., & Banjevic, D. (2006). A review on machinery diagnostics and prognostics implementing condition-based maintenance. *Mechanical Systems and Signal Processing*, 20(7), 1483–1510.
Kusiak, A., et al. (2013). Maintenance optimization of wind turbines using data-driven approaches. *Renewable Energy*, 45, 11–16.
Na, J., et al. (2020). Predictive maintenance for water management using artificial intelligence: A systematic review. *Journal of Environmental Management*, 258, 110038.
Nguyen, T., et al. (2022). Artificial intelligence in predictive maintenance: A bibliometric analysis and research agenda. *Journal of Manufacturing Technology Management*, 33(3), 480–499.
Olejnik, S., et al. (2020). The importance of AI-based maintenance in critical infrastructure. *Sustainability*, 12(23), 10092.
Ren, X., et al. (2021). Predictive maintenance using deep learning: A case study in wind energy. *IEEE Transactions on Industrial Informatics*, 17(8), 5486–5495.
Schwabacher, M., & Goebel, K. (2007). A survey of artificial intelligence for prognostics. *AAAI Fall Symposium: AI for Prognostics*, 1–8.
Van Thienen, P., et al. (2020). AI-based predictive maintenance for water distribution systems: Challenges and opportunities. *Water Research*, 170, 115353.
Zhang, Q., et al. (2019). Deep learning for predictive maintenance of industrial equipment: A review. *IEEE Access*, 7, 62624–62634.
Zhao, Z., et al. (2020). An artificial intelligence approach to predictive maintenance for telecommunications. *Journal of Telecommunications and Information Technology*, 2, 100–111.
Zhou, X., et al. (2022). AI-enabled predictive maintenance for smart energy grids: A review. *Energy Reports*, 8, 3597–3610.
Zio, E. (2013). Prognostics and health management of industrial equipment: A review. *Annual Reviews in Control*, 37(1), 1–16.
Zonta, T., et al. (2020). Predictive maintenance in the Industry 4.0: A systematic literature review. *Computers & Industrial Engineering*, 150, 106889.

9 Cluster Analysis
A Tool for Technology Management

THE VALUE OF CLASSIFICATION

Classifying objects into similar groups is a task most of us would have encountered in our daily lives (Ray and Aiyappan, 2014). A child classifies his toys according to his favorite colors and a real estate agent classifies his homes according to geography and price. A direct marketer classifies his target according to geographic, demographic, psychographic, and behavioral attributes. In the business world, the importance of market segmentation is growing. In the context of this book, cluster analysis is useful for classifying technology. With the advent of technology that gives access to data mining and manipulation, analysts can customize their market segments appropriately. Access to desktop computers to solve business problems using statistical analytical tools is gaining importance and priority. The search for classifications of objects or persons is common to many disciplines. As the field of classification, science became more sophisticated, and the emphasis grew to rely more on objective techniques of numerical taxonomy. The advent of high-speed desktop computers that have enormous storage capacities greatly facilitated the application of advanced statistical tools in business situations. The common term used for the class of procedures that are used to isolate the component data into groups is cluster analysis. Application of cluster analysis is prevalent in diverse areas as (a) psychology for classifying individuals into personality types; (b) chemistry for classifying compounds as per their properties; (c) regional analysis for classifying cities as per their demographic and other variables; (d) marketing analysis for classifying customers into segments based on their buying behavior and product use.

Cluster analysis has become a standard tool for marketing research applications. Despite its frequent use, there is a general lack of knowledge regarding the clustering methodology, as is apparent from frequent omission in the literature, about the specification of the clustering method used in a case. In the past, cluster analysis has been viewed with skepticism (Frank and Green, 1968; Inglis and Johnson, 1970; Wells, 1975). This skepticism probably arose due to the confusing array of names like typology, classification analysis, numerical taxonomy, and Q-analysis for this technique. The names differ across disciplines, but they all deal with classification according to some natural relationships. This chapter presents the multivariate statistical techniques that are popular in providing business solutions and specifically focuses on a

DOI: 10.1201/9781032659220-9

clustering methodology called K-means clustering and illustrates the use of this type of statistical technique in a marketing situation.

MULTIVARIATE ANALYSIS

Multivariate analysis examines the statistical properties among three or more variables across multiple subjects. The multivariate statistical techniques that are suitable for segmentation purposes consist of *dependence* and *interdependence* techniques. *Dependence* techniques use one or more types of independent variables to predict or explain a dependent variable. The commonly used *dependence* techniques for segmentation research include the following (Myers, 1996): automatic interaction detector (AID), Chi-square automatic interaction detector (CHAID), and regression and discriminant analysis. *Interdependence* techniques search for groups of people or items that are found to be similar in terms of one or more sets of basic variables. All variables used are considered to be more or less equal in terms of interest. The most commonly used interdependence techniques for segmentation research include the following: hierarchical clustering, partition clustering, and Q-type factor analysis. The conceptual framework of multivariate statistical methods is presented by Doyle (1977). The framework appears in three levels, summarized below:

First level:
- All multivariate methods

Second level:
- Dependent methods
  - AID
  - CHAID
  - Regression and discriminant analysis
- Interdependent methods.
  - Hierarchical clustering
  - Partition clustering
  - Q-type factor analysis

The multivariate analytical techniques that are often used to provide business solutions include the following:

- Cluster analysis
- Multiple regression methods
- Multiple discriminant analysis
- Multiple analysis of variance (MANOVA)
- Canonical correlation
- Linear probability models (LOGIT, PROBIT)
- Conjoint analysis
- Structural equation modeling
- Factor analysis
- Multidimensional scaling
- Correspondence analysis

MULTIVARIATE ANALYSIS IN CLUSTERING

A multivariate analysis is a step beyond univariate analysis and utilizes more than one variable at the same time to explain and divide up the population under study. Variables used as input can be of several forms, and cluster analysis is often that further step from a univariate classification (TargetPro Version 4.5, 2003). Prepackaged approaches to multivariate statistical clustering use customized versions of non-hierarchical cluster analysis, known variously as "iterative centroidal relocation" or "K-means clustering." This approach adjusts in multidimensional space the definition of a fixed number of clusters until a criterion involving "sums of squared distances" is minimized. The computer tests a number of different classifications and searches for a set that maximizes the similarity of the objects assigned to the same cluster and, at the same time, maximizes the statistical distances or differences between individual clusters.

MULTIVARIABLE DATA MANAGEMENT IN SEGMENTATION

Generally, two approaches are used to reduce the amount of data to be analyzed. The first approach is to select on an a priori basis a relatively small number of variables (40–50) to represent each dimension that is more important. The problem with this approach is that a lot of subtlety hidden in the remaining variables is lost. The second approach is called principal component analysis, a statistical process that groups individual variables into separate components of factors and uses these, rather than individual variables as a basis for measuring similarities between areas (Wind, Y., 1978). This is an excellent technique for removing distortion caused by taking too many variables from one domain and not enough from another. The disadvantage of using principal components analysis is that it is less effective than using individual variables for building classification systems, the level of customization is limited and does not represent an individualistic approach of the end customer.

CLUSTER ANALYSIS

In several statistical procedures, the objects on which measurements are made are assumed to be homogeneous. However, in cluster analysis, the focus is on the possibility of dividing a set of objects into several subsets of objects that display systematic differences. Cluster analysis represents a group of multivariate techniques used primarily to identify similar entities based on the characteristics they possess. It identifies objects that are very similar to other objects in groups or "clusters" with respect to some predetermined selection criteria, while the groups exhibit significant differences between them regarding the same criteria. The object cluster then shows high within-cluster homogeneity and high between-cluster heterogeneity. Figure 9.1 illustrates the within-cluster-and-between-cluster variations

There is no guideline on what constitutes a cluster, but ultimately, it depends on the value judgment of the user. The data are allowed to determine the pattern of clusters inherent in the data. In the literature, an almost endless number of clustering algorithms are found. All of these depend upon high-speed computer capacity and aim to meet some criteria that maximize the between-cluster variation compared to

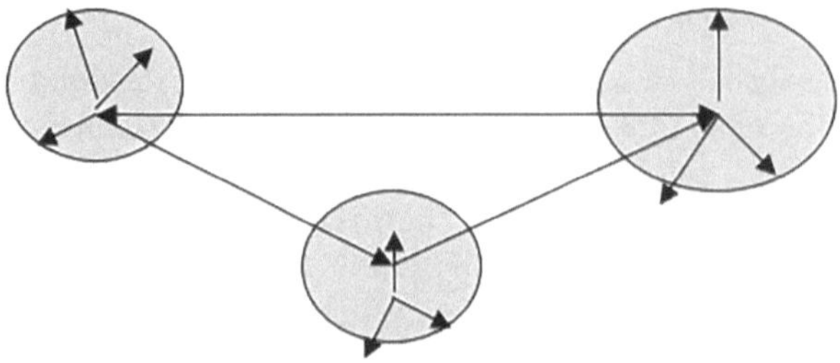

FIGURE 9.1 Within-Cluster and Between-Cluster Variations.

the within-cluster variation. Many different approaches have been used to measure inter-object similarity and numerous algorithms to develop clusters. But till now, there is no universally agreed upon definitive answer to these methods. As such, cluster analysis remains much more of an art than an exact science.

PHASES OF CLUSTERING ANALYSIS

The process of cluster analysis may be divided into three phases (Hair, et al., 1987). These are:

- Partitioning,
- Interpretation, and
- Profiling.

Partitioning

During the first phase, an appropriate measure is selected for measuring inter-object similarity. The proximity or closeness between each pair of objects is used as a measure of similarity. Since distance is a complement of similarity, it is used as a measure of similarity.

Distance type measurement: this type of measurement is possible for quantitative data. The general Minkowski metric for distance measurement is defined by equation 9.1

$$d_{ij} = \left\{ \sum_{k=1}^{p} \left| X_{ik} - X_{ij} \right|^{r} \right\}^{1/r} \tag{9.1}$$

where d_{ij} is the distance between two objects i and j.

When r = 2, the Minkowski equation reduces to the familiar Euclidean distance between the objects i and j and is given by equation 9.2.

$$d_{ij} = \left\{ \sum_{k=1}^{p} \left(X_{ik} - X_{jk} \right)^{2} \right\}^{1/2} \tag{9.2}$$

When r = 1, the Minkowski equation reduces to the city-block metric, which is given by equation 9.3.

$$d_{ij} = \sum_{k=1}^{p} \left| X_{ik} - X_{jk} \right| \tag{9.3}$$

Several other options are available in various computer programs. One option is to use the sum of squared differences as a measure of similarity.

The raw data are converted to Z-scores before computing distances. This step is taken to eliminate the spurious effect of unequal variances of the variables. Another very useful Euclidean distance measure is the Mahalanobis distance. The Mahalanobis D^2 generalized distance measure is comparable to R^2 in regression analysis and is superior to the different versions of the Euclidean distance measures. It is given by equation 9.4.

$$(Xi - Xj)' S - 1(Xi - Xj) \tag{9.4}$$

where S is the pooled within-group covariance matrix, and Xi and Xj are the respective vectors of measurements on objects i and j. This distance measurement has the advantage of explicitly accounting for any correlations that might exist between the variables.

Match type measurement: for qualitative data, a match type or association measure is suitable. The association generally takes the value of zero to indicate the absence of an attribute and 1 to indicate the presence of the attribute. Two objects or individuals are considered similar if they share the same common attributes and dissimilar if they do not share the common attributes. We can visualize the variables' absence (0) and presence (1) in Table 9.1.

Similarity may be measured by counting the total number of matches (either 0, 0) or (1, 1) between X and Y and dividing the total by the number of attributes (8). The similarity between X and Y then in this case is given by:

$$\text{Similarity between X and Y} = \left\{ \left(\left(\text{Number of } (1,1) \text{s or } (0,0) \text{s} \right) / \text{No.of attributes} \right\}^{*} \; 100\% = (5/8)^{*} \, 100\% = 62.5\%. \tag{9.5}$$

The resulting association table is given by Table 9.2.

TABLE 9.1
Contingency table of similarity

Object	1	2	3	4	5	6	7	8
X	0	1	1	0	1	1	1	1
Y	1	1	1	0	0	1	0	1

TABLE 9.2
Association table for X and Y

	Object 2		Total
	+	–	
+ Object 1-	4	2	62
	1	1	
Total	5	3	8

Association measures can be computed in several different ways, and unfortunately, these result in different values for the same data sets (Dillon, et al. 1984). Hence, it is essential to assign 1's and 0's based on the importance to the user.

Clustering algorithms: The next step is to select a particular type of computational algorithm. The commonly used clustering algorithms are of two types: hierarchical and non-hierarchical. The clustering process strives to maximize between-cluster variability and minimize within-cluster variability. In other words, subjects within the cluster are most similar, and each cluster is markedly different from the others. Clustering techniques have been applied to a wide variety of research problems. For example, in the field of medicine, clustering is very useful for developing taxonomies of diseases (Hartigan, J., et al., 1979). In the field of psychiatry, the correct diagnosis of clusters of symptoms such as paranoia and schizophrenia is essential for successful therapy. In general, whenever one needs to classify a large number of information into manageable piles, cluster analysis is of great utility.

Hierarchical clustering: Hierarchical procedures construct tree-like structures. There are basically two types of procedures: agglomerative and divisive. In the agglomerative method, all cases start in their own cluster and are then combined into smaller and smaller numbers of clusters. In other words, all cases start off in the same cluster, and the process commences by dividing the customers into two groups. The group with the most internal variation, the least homogeneous, gets split in two, and now, there are three groups, and so on, and the process continues until it can no longer find a statistical justification to continue (Hartigan, J., et al., 1979).

NON-HIERARCHICAL CLUSTERING K-MEANS CLUSTERING METHOD

This technique of clustering is gaining popularity for large databases and can be used once an agreement is reached with regard to the number of clusters. The non-hierarchical procedure does not involve a tree-like construction process but needs to select a cluster center or seed, and all objects within a specified distance are included in the resulting cluster. There are three different approaches for non-hierarchical clustering based on sequential threshold, parallel threshold, or optimizing procedures.

The K-means clustering splits a set of objects into a selected number of groups by maximizing between variations relative to within variation (Green and Rao, 1969).

In general, the K-means method will produce exactly k different clusters of greatest possible distinction (Sherman and Seth, 1977; Ling and Li, 1998). In the K-means clustering procedure, the value of K or the number of clusters has to be decided prior to processing. There appears to be no standard method, but some guidelines are available (Hair et al., 1987). The clustering process may be stopped when the distance between clusters at successive steps exceeds a pre-selected value. An intuitive number of clusters may be tried, and based on some pre-selected criteria, the best among the alternatives may be selected. Frequently, the judgment of practicality regarding comprehension and communication becomes the deciding factor.

Cluster analysis packages may be purchased as "off-the-shelf" software that uses segmentation techniques based on neighborhood type approach (TargetPro Version 4.5, 2003). These software use prepackaged approaches to multivariate statistical clustering which fundamentally follow the same concept. Several of these packages use a customized version of non-hierarchical cluster analysis, known as "K-means clustering." This approach consists of testing a number of different classifications and searching for a set of clusters that maximizes the similarity of all the geographic units assigned to the same cluster and, at the same time, maximizes the statistical differences between individual clusters.

INTERPRETATION PHASE

This phase involves determining the nature of the clusters by examining the criteria used to develop the clusters. One way is to determine the average value of the objects in each cluster for each raw variable and develop average profiles from these data. A cluster may favor one attitude, while the other may favor another. From this analysis, each cluster's attitudes may be evaluated and significant interpretations developed. The interpretations facilitate to assign label that represents the nature of the clusters.

PROFILING PHASE

This phase involves describing the characteristics of each cluster to explain the way they differ on relevant dimensions. The demographics, behavioral patterns, buying habits, consumption characteristics, and other traits relevant to a particular study are usually included in the analysis for profiling. For example, more affluent and younger customers may represent one cluster while the other one may represent older and more conservative persons. This analysis has to focus on the inherent characteristics that differ significantly from those of the other clusters and are different from the ones used to develop the clusters.

TESTING VALIDITY OF CLUSTERING SOLUTION

Assessment consists of examining the following:

- Distinctiveness of clusters, presented by profiling,
- Optimum number of clusters depending on a balance between the extent of homogeneity within the cluster and the number of clusters, and

- Goodness of fit indicated by high rank-order correlation between the input and the solution output. Because clusters are generated by maximizing the between-cluster sums-of-squares, the usual test of significance of ANOVA ($F\ \alpha, \nu 1, \nu_2$) cannot be conducted in the case of cluster analysis (Dillon et al., 1984). Instead, the maximum value of the F-statistic among the different alternative groupings is used as an indication of best fit.

APPLICATION OF CLUSTER ANALYSIS IN MARKETING MANAGEMENT

A company's market area is composed of several segments of users, as defined by any number of factors, including industry type, company size, purchase and usage behavior, demographic, psychographics, and sociographic factors. In order to maintain its competitive advantage, the company has to target its market and streamline its communication strategies. Segmentation is a marketing technique that facilitates targeting a group of customers with specific characteristics (Keller, W.J., 1993). By segmenting the market, a company can target product offerings and communications to the appropriate segments for higher overall effectiveness and profitability. Market penetration is defined as the percentage of the available target market currently served by the company (Kotler, P., 1992). Over the past 30 years, encouraged by developments in microchip technology and mass storage devices, the use of computer technology in business applications has expanded rapidly (Freeman, P., 1991). This trend has permeated into marketing data analysis, and a combination of computing power and statistical concepts has played a vital role in advancing the field of marketing management for dealing with a large number of variables. Marketers have to treat different groups or segments differently to derive maximum effectiveness. Cluster analysis is a technique that helps to identify relatively homogenous groups of objects on selected characteristics, using an algorithm that can deal with a large number of objects (Kaufman and Rousseeuw, 1990). Cluster analysis has been traditionally seen as a complex technique, largely in the domain of mathematicians (Blamires, C., 1995). With the advent of mass computing and data storage systems, this technique has now become practical in market segmenting. There are several procedures for clustering, and selection has to be made based on an individual case.

OBJECTIVES OF THE STUDY

The marketing team of a Fortune 500 corporation with customers all over the continental United States wanted to analyze and segment the designated marketing areas (DMAs) that it serves into specific "clusters" that behaved similarly to each other. They also had the additional objective of understanding which and how the multiple variables affected market penetration among its segments (Ray, et al., 2005).

METHODOLOGY

The methodology of this research consisted of:

- Data reduction by principal components analysis, another statistical technique
- Market segmentation by K-means cluster analysis of the DMAs based on each variable and penetration using SPSS 11.5 for Windows software, and

- Analysis of variance (ANOVA) to test the significance of difference between group means for each variable on the DMA penetration.

VARIABLES

Principal component analysis was conducted to reduce 400 data variables into 29 to make the analysis pragmatic and actionable in the marketing sense. Out of these variables, 18 had usable data in terms of how recently the data was collected, the source, and the validity of the data. These 18 data variables were:

- Total media spending
- Total amount spent
- All circulation
- Radio April 2002
- Newspaper
- April roll fold amount
- Number of centers
- Number of travelers
- Median home value
- Average household income
- Percentage urban
- Percentage 55–64
- Percentage spent on rent
- Percentage never married
- Percentage with two incomes
- Average liquor
- Average pizza
- Average items.

DATA COLLECTION, STORAGE, ACCESS, AND FORMATS

The variables were reduced using the principal components analysis to factor the hundreds of variables into meaningful variables. The data was obtained in a Visual FoxPro (dBase IV) format. The original file used for the principal components analysis was data from a leading rental agency that appended information (e.g., demographical and socio-economical) into the customer file. The dBase file was imported into the SPSS software and stored as an SPSS (.SAV) file.

PARTITIONING – ALLOCATION OF DMAs TO CLUSTERS USING K-MEANS CLUSTERING METHODOLOGY

The statistical package SPSS 11.5 for Windows could compute clusters using the K-means clustering methodology (Marija, 2002). The requirement for this was to know the variables that needed to be used for creating the SPSS syntax to group DMAs. Each variable was used along with the penetration to create the clusters. The syntax instructs the software to choose a variable and to iterate and classify the cases based on DMA into five clusters that are most disparate based on the principle of maximizing between-group variance and minimizing within-group variance. The number of iterations is set to a maximum of ten which is a reasonable number to find the cluster solution. The input also requires the determination of the number of clusters for each case and was determined based on heuristics so that the clusters created could be used for marketing action at a later stage. The syntax also asks to display the cluster membership for each DMA and output the ANOVA table. Each DMA is output as a member belonging to a specific cluster.

RESULTS: INTERPRETATION AND PROFILING OF CLUSTERS

Based on this study, the DMAs were grouped into clusters with significant differences for eight of the 18 variables. The remaining ten variables did not have significant differences across the DMAs and were not instrumental in the penetration difference as indicated by the lower F-values in the analysis of variance (ANOVA). See Table 9.3.

Thus, these could not be used for creating valid clusters which behaved similarly within groups and differently across groups with respect to the variable and penetration index. The significant eight variables were:

- Total media spending
- Total amount spent
- All circulation
- April roll fold amount
- Number of centers
- Number of travelers
- Median home value
- Average household income

The resulting impact for each of the eight significant variables was determined individually. Each report contained the final cluster centers, the ANOVA table which indicated the F-values (higher F-values (typically more than 1.0) indicated higher differences in variables across DMAs), the number of DMAs in each cluster, and the cluster membership details. The results for the variable "Number of Centers" is given in Table 9.4.

TABLE 9.3
Variable: number of centers

Cluster	Number of DMAs	Penetration Index	No. Centers Index
1	1	8	2837
2	52	93	51
3	7	101	183
4	42	148	82

TABLE 9.4
F-values for Significant Variable Index and Penetration Index

No.	Significant Variable	Variable Index F-value	Penetration IndexF-value
1.	Total media spending	562	33
2.	Total amount spent	23,823	42
3.	All circulation	105	31
4.	April roll fold amount	363	39
5.	Number of centers	3916	40
6.	Number of travelers	1930	62
7.	Median home value	206	100
8.	All circulation	105	31

Results for each cluster and representative DMAs were provided for a better understanding of the analysis by the client. An example of the results for the variable "Number of Centers" is given below:

- Cluster 1 was an outlier and has only one DMA. There were three centers for a core population of 3223 giving it a high index (2837).
- Cluster 2 had most of the low-penetrated DMAs from all regions and the lowest number of centers per core population for any cluster. This fact indicated that increasing the number of centers in these DMAs provided the potential for improving penetration.
- Cluster 3 has DMAs where it seemed that increasing centers would not prove as effective because these DMAs typically had a high number of centers per core population.
- Cluster 4 had the majority of high-performing DMAs from all regions. These DMAs had high penetration and high index of centers but not as high as cluster 3 which suggests that there could be more additions of centers to improve penetration.
- Interestingly, when analyzing a few larger DMAs like Atlanta, Chicago, Houston, and Denver, they were found to be segmented in Cluster 2 which had a low index for centers and a low penetration with this variable. These could be potential markets where increasing the number of centers would improve penetration. The centers were not well represented in these large DMAs.
- It was interesting to note that other large DMAs such as Charlotte, Orlando, and San Francisco fell in cluster 4 which is representative of DMAs which had a higher number of centers and a high penetration index with respect to the variable. The output from ANOVA verified that the variable "Number of Centers" had a significant effect on penetration.

TESTING VALIDITY OF CLUSTERING SOLUTION USING ANOVA

The hypothesis to test is as follows:

$$H_0 : \mu_1 = \mu_2 = \mu_3 = \ldots.. = \mu_r \tag{9.6}$$

$$H_1 = \text{Not all are equal}\left(i = 1\ldots\ldots\ldots r\right). \tag{9.7}$$

There are r populations or treatments (DMS in this case). The null hypothesis assumes that all the means across the various DMSs are the same for each variable. The alternate hypothesis assumes that at least one DMA differs from the available DMAs in terms of the variables analyzed.

SPSS 11.5 for Windows was used to compute the ANOVA test statistic or F-value for the difference in the means of the variables across each segment or cluster. A high F-value (typically greater than 1) indicates that the means of the variables vary considerably across the clusters and hence is a factor affecting penetration across DMAs. The ANOVA table showed F-values for each of the eight significant variables and market penetration indexes. These are given in Table 9.4.

Since the F-values for all variables are much greater than 1, we conclude that the four clusters are significantly different from each other with respect to the variables and penetration indexes. These high values of F-statistic were achieved, as intended to enhance, by the clustering technique.

The F-tests were used only for descriptive purposes because the clusters have been chosen to maximize the difference among cases in different clusters. The observed significance levels are not corrected for this and thus cannot be interpreted as tests of the hypothesis that the cluster means are equal. High values of the test static F indicate significant differences between the clusters. The variables were assumed to be normally distributed for conducting ANOVA. This assumption was tested by histogram for each variable and was found to be acceptable, although in some cases, there was a slight skew to the left or right.

LIMITATIONS OF THE CASE STUDY

Due to the inherent data insufficiencies, 11 factored variables were dropped, and the remaining 18 were used for this study. These 18 variables were assumed to be representative of the initial data elements and were used for determining the significant ones that affect penetration. All underlying data were for the year 2001 only.

REFERENCES

Blamires, Chris. (1995). Segmentation techniques in Market research: Exploding the mystique surrounding cluster analysis-part 1. *Journal of Targeting, Measurement and Analysis for Marketing*, Vol. 3, pp. 338.

Doyle, P. (1977). In Marketing: A Review. *Journal of Business Research*, Vol. 5 (September), pp. 235–248.

Frank, R.E., and Green, P.E. (1968). Numerical Taxonomy in Marketing Analysis: A Review Article. *Journal of Marketing Research*, Vol. 5 (February), pp. 83–93.

Freeman P. (1991). Using computers to extend analysis and reduce data. *Journal of Market Research Society*, Vol. 33 (2), pp. 127–136.

Green, Paul E. and Rao, Vitala R., (1969). A Note on Proximity Measures and Cluster Analysis. *Journal of Marketing Research*, Vol. 6, pp. 359–364.

Hair, J.F., Anderson, R.E., and Tatham, R.L. (1987). *Multivariate Data Analysis*. Macmillan Publishing Company, New York.

Inglis, J., and Johnson, D. (1970). Some Observations On and Developments In, the Analysis of Multivariate Survey Data. *Journal of the Market Research Society*, Vol. 12, pp. 75–80.
Kaufman, L., and Rousseeuw, P.J. (1990). *Finding Groups in Data: An Introduction to Cluster Analysis*. John Wiley & Sons, New York.
Keller, W.J. (1993). Trends in survey data processing, *Journal of the Market Research Society*, Vol. 35 (3), pp. 211–219.
Kotler, Philip. (1992). *Marketing Management*. 8th ed., Englewood Cliffs, NJ, Prentice-Hall.
Ling, C. X. and Li, C.I. (1998). Data Mining for Marketing: Problems and Solutions, *Proceedings of the 4th International Conference on Knowledge Discovery & Data Mining*, pp. 73–79.
Marija, J. N. (2002).SPSS@11.0, *Guide to Data Analysis*. Prentice Hall Inc, New Jersey.
Myers, James H. (1996). *Segmentation and Positioning for Strategic Marketing Decisions*. American Marketing Association. pp. 30–32.
Ray, Paul S. and Aiyappan, H. (2014), "Cluster Analysis: A Tool for Industrial Engineers," Chapter 9 in *Handbook of Industrial and Systems Engineering*, 2 ed, Adedeji Badiru, Editor, Taylor & Francis Group/CRC Press, Boca Raton, FL, 2014.
Ray, P.S., Aiyappan, H., Elam, M.E. and Merritt, T.W. (2005). *International Journal of Industrial Engineering*, Vol. 12 (2), pp. 125–131.
Sherman, L. and Seth, Jagdish N. (1977). Cluster Analysis and its Applications in Marketing Research. In *Multivariate Methods for Market and Survey Research*, Jadish N. Seth (ed.), Chicago, pp. 193–208. American Marketing Association,
TargetPro Version 4.5. (2003). Users Guide- MapInfo Corporation.
Wells, W.D. (1975). Psychographics: A Critical Review. *Journal of Marketing Research*, Vol. 12 (May), pp.196–213.
Wind, Yoram. (1978). Issues and Advances in Segmentation Research, *Journal of Marketing Research*, Vol. 15, pp. 317–337.

10 DEJI Systems Model
A Tool for Industrial Engineering

TECHNOLOGY, EFFICIENCY, AND QUALITY

Systems quality is at the intersection of efficiency, effectiveness, and productivity. Efficiency provides the framework for quality in terms of resources and inputs required to achieve the desired level of quality. Effectiveness comes into play with respect to the application of product quality to meet the specific needs and requirements of an organization. Productivity is an essential factor in the pursuit of quality as it relates to the throughput of a production system. To achieve the desired levels of quality, efficiency, effectiveness, and productivity, a new research framework must be adopted. In this chapter, we present a quality enhancement model for quality design, evaluation, justification, and integration (DEJI) based on a product development application. The model is relevant for research efforts in quality engineering and technology applications and other systems engineering applications, particularly in educational programs (Miller, 2019).

The DEJI model of systems engineering provides one additional option for systems engineering development applications. Although the model is generally applicable in all types of systems modeling, systems quality is specifically used to describe how the DEJI model is applied. The core stages of the DEJI model are:

- Design
- Evaluation
- Justification
- Integration

Design encompasses any system initiative providing a starting point for a project. Thus, design can include technical product design, process initiation, and concept development. In essence, we can say that "design" represents requirements and specifications. Evaluation can use a variety of metrics, both qualitative and quantitative, depending on the organization's needs. Justification can be done based on monetary, technical, or social reasons. Integration needs to be done with respect to the normal or standard operations of the organization. Figure 10.1 illustrates the full profile of the DEJI model.

This chapter advocates using the DEJI systems model for technology and systems integration. An application guide is presented in a structural template for systems design, evaluation, justification, and integration. Systems quality is used as the basis for illustrating how the model is applied across its four stages. Specific emphasis is

DOI: 10.1201/9781032659220-10

FIGURE 10.1 The DEJI Systems Model.

placed on applying the model to technology management in the context of pursuing and tackling UN's SDGs and NAE's grand challenges.

All the operational elements embedded in the DEJI model are explained and described as presented below:

Design embodies agility, define end goal, and engage stakeholder.
Evaluate embodies feasibility, metrics, gather evidence, and assess utility.
Justify embodies desirability, focus on implementation, and articulate conclusions.
Integrate embodies affordability, sustainability, and practicality.
For application purposes, these elements interface and interact systematically to enhance the overall operational performance of the organization.

APPLICATION TO TECHNOLOGY QUALITY SYSTEM

Several aspects of quality must undergo rigorous research along with the realms of both quantitative and qualitative characteristics. Many times, quality is taken for granted, and the flaws only come out during the implementation stage, which may be too late to rectify. The growing trend in product recalls is a symptom of a priori analysis of the sources and implications of quality at the product conception stage. This column advocates the use of the DEJI model for enhancing quality design, quality evaluation, quality justification, and quality integration through hierarchical and stage-by-stage processes.

Better quality is achievable, and there is always room for improvement in the quality of products and services. But we must commit more efforts to the research at the outset of the product development cycle. Even the human elements of the perception of quality can benefit from more directed research from a social and behavioral sciences point of view.

TECHNOLOGY ACCOUNTABILITY

Throughout history, engineering has answered the call of the society to address specific challenges (Grayson, 1993; Kirby et al., 1990). With such answers comes a greater expectation of professional accountability. Consider the level of social responsibility that existed during the time of the Code of Hammurabi (Bryant, 2005). Two of the laws are echoed below:

> **Hammurabi's Law 229:**
> If a builder build a house for someone, and does not construct it properly, and the house which he built fall in and kill its owner, then that builder shall be put to death.
>
> **Hammurabi's Law 230:**
> If it kills the son of the owner the son of that builder shall be put to death.

These are drastic measures designed to curb professional dereliction of duty and enforce social responsibility with a particular focus on product quality. Research and education must play bigger and more direct roles in the design, practice, and management of quality. Duckworth and Moore (2010) present modern aspects of social responsibility in the context of day-to-day personal and professional activities. The global responsibility of the greater society is presented by Bhargava (2006) with respect to world development challenges covering the global economy, human development, global governance, and social relationships. Quality is the common theme in the development challenges. Focusing on the emerging field of Big Data, Xian and Madhavan (2014) advocate engineering education collaboration, which aligns well with data-intensive product development. With the above principles as possible tenets for better research, education, and practice of quality in engineering and technology, this column presents the DEJI model as a potential methodology.

THE DEJI MODEL

The DEJI model encourages the practice of building quality into a product right from the beginning so that the product integration stage can be more successful. Figure 10.2 shows the four-legged model for an overall systems quality with respect to design, evaluation, justification, and integration. Table 10.1 shows the potential elements within each stage of the DEJI model.

THE DESIGN PHASE

The design of quality in product development should be structured to follow point-to-point transformations. A good technique to accomplish this is the use of state-space transformation, with which we can track the evolution of a product from the concept stage to the final product stage.

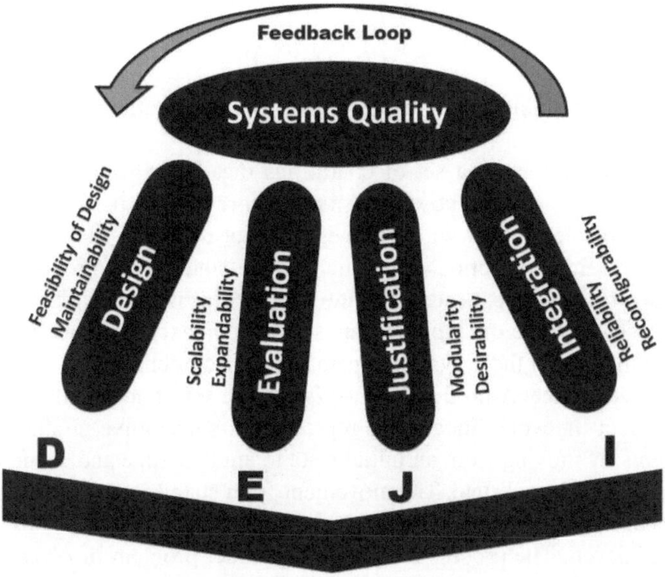

FIGURE 10.2 Footstool Elements of the DEJI Systems Model.

TABLE 10.1
Tabulated Layout of DEJI Model

DEJI Model	Characteristics	Tools & Techniques
Design	Define goalsSet performance metricsIdentify milestones	Parametric assessmentProject state transitionValue stream analysis
Evaluate	Measure parametersAssess attributesBenchmark results	Pareto distributionLife cycle analysisRisk assessment
Justify	Assess economicsAssess technical outputAlign with goals	Benefit-cost ratioPayback periodPresent value
Integrate	Embed in normal operationVerify symbiosisLeverage synergy	SMART conceptProcess improvementQuality control

Design, in the context of the DEJI systems model, is not limited just to the physical design of a product. Design encompasses a collaborative template across multiple factors, including the following:

- Legal considerations
- Regulatory requirements
- Safety concerns
- Reliability expectations
- Sustainability mandate
- Environmental impact
- Lifecycle cost reduction
- Optimization across platforms

- Maintainability
- Human-system integration

For the purpose of technology quality design, the following definitions are applicable:

Product state: A state is a set of conditions that describe the product at a specified point in time. The *state* of a product refers to a performance characteristic of the product which relates an input to output such that knowledge of the input function over time and the state of the product at time $t = t_0$ determines the expected output for $t \geq t_0$. This is particularly important for assessing where the product stands in the context of new technological developments and the prevailing operating environment.

Product state-space: A product *state-space* is the set of all possible states of the product lifecycle. State-space representation can solve product design problems by moving from an initial state to another state and eventually to the desired end-goal state. The movement from state to state is achieved by means of actions. A goal is a description of an intended state that has not yet been achieved. The process of solving a product problem involves finding a sequence of actions that represents a solution path from the initial state to the goal state. A state-space model comprises state variables that describe the prevailing condition of the product. The state variables are related to inputs by mathematical relationships. Examples of potential product state variables include schedule, output quality, cost, due date, resource, resource utilization, operational efficiency, productivity throughput, and technology alignment. For a product described by a system of components, the state-space representation can follow the quantitative metric below:

$$Z = f(z, x); \quad Y = g(z, x),$$

where f and g are vector-valued functions. The variable Y is the output vector while the variable x denotes the inputs. The state vector Z is an intermediate vector relating x to y. In generic terms, a product is transformed from one state to another by a driving function that produces a transitional relationship given by:

$$S_s = f(x \mid S_p) + e,$$

where S_s = subsequent state; x = state variable; S_p = the preceding state; e = error component

The function f is composed of a given action (or a set of actions) applied to the product. Each intermediate state may represent a significant milestone in the project. Thus, a descriptive state-space model facilitates an analysis of what actions to apply in order to achieve the next desired product state. Figure 10.3 shows a representation of a product development example involving the transformation of a product from one state to another through the application of human or machine actions. This simple representation can be expanded to cover several components within the

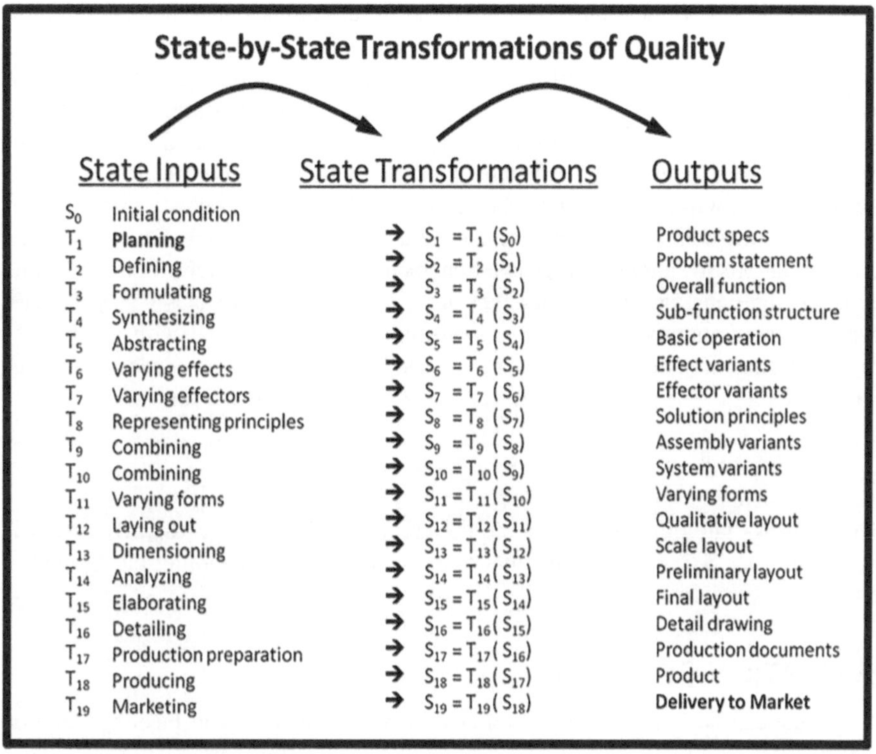

FIGURE 10.3 State-Space Transformations of Quality.

product information framework. The hierarchical linking of product elements provides an enhanced transformation structure. The product state can be expanded in accordance with implicit requirements. These requirements might include grouping of design elements, linking precedence requirements (both technical and procedural), adapting to new technology developments, following required communication links, and accomplishing reporting requirements. The actions to be taken at each state depend on the prevailing conditions of the product. The nature of subsequent alternate states depends on the actions implemented. Sometimes, multiple paths lead to the desired end result. At other times, only one unique path can lead to the desired objective. In conventional practice, the characteristics of the future states can only be recognized after the fact, thus making it impossible to develop adaptive plans. In the implementation of the **DEJI** model, adaptive plans can be achieved because the events occurring within and outside the product state boundaries can be taken into account.

If we describe a product by P state variables s_i, then the composite state of the product at any given time can be represented by a vector **S** containing P elements. That is,

$$\mathbf{S} = \{s_1, s_2, \ldots, s_P\}$$

The components of the state vector could represent either quantitative or qualitative variables (e.g., cost, energy, color, time). We can visualize every state vector as a point in the state-space of the product. The representation is unique because every state vector corresponds to one, and only one point in the state-space. Suppose we have a set of actions (transformation agents) that can be applied to the product information to change it from one state to another within the project state-space. The transformation will change a state vector into another state vector. A transformation may be a change in raw material or a change in design approach. The number of transformations available for a product characteristic may be either finite or unlimited. We can construct trajectories that describe the potential states of a product's evolution as we apply successive transformations in accordance with technology forecasts. Each transformation may be repeated as many times as needed. Given an initial state $\mathbf{S}_0$, the sequence of state vectors is represented by the following:

$$\mathbf{S}_n = T_n\left(\mathbf{S}_{n-1}\right).$$

The state-by-state transformations are then represented as $\mathbf{S}_1 = T_1(\mathbf{S}_0)$; $\mathbf{S}_2 = T_2(\mathbf{S}_1)$; $\mathbf{S}_3 = T_3(\mathbf{S}_2)$; **...**; $\mathbf{S}_n = T_n(\mathbf{S}_{n-1})$. The final state, $\mathbf{S}_n$, depends on the initial state **S** and the effects of the actions applied.

EVALUATION OF QUALITY

A product can be evaluated based on cost, quality, schedule, and meeting requirements. Many quantitative metrics can be used in evaluating a product at this stage. Learning curve productivity is one relevant technique that can be used because it offers an evaluation basis of a product with respect to the concept of growth and decay. The half-life extension (Badiru, 2012) of the basic learning is directly applicable because the half-life of the technologies going into a product can be considered. In today's technology-based operations, retention of learning may be threatened by fast-paced shifts in operating requirements. Thus, it is of interest to evaluate the half-life properties of new technologies as they impact the overall product quality. Information about the half-life can tell us something about the sustainability of learning-induced technology performance. This is particularly useful for designing products whose life cycles stretch into the future in a high-tech environment.

JUSTIFICATION OF QUALITY

We need to justify a program based on quantitative value assessment. The systems value model (SVM) is an effective quantitative technique that can be used here for project justification based on value. The model provides a heuristic decision aid for comparing project alternatives. It is presented here again for the present context. Value is represented as a deterministic vector function that indicates the value of tangible and intangible attributes that characterize the project. It is represented as $V = f\left(A_1, A_2, \ldots, A_p\right)$, where V is the assessed value and the A values represent quantitative measures or attributes. Examples of product attributes are quality, throughput, manufacturability, capability, modularity, reliability, interchangeability, efficiency, and cost performance. Attributes are considered to be a combined function of factors.

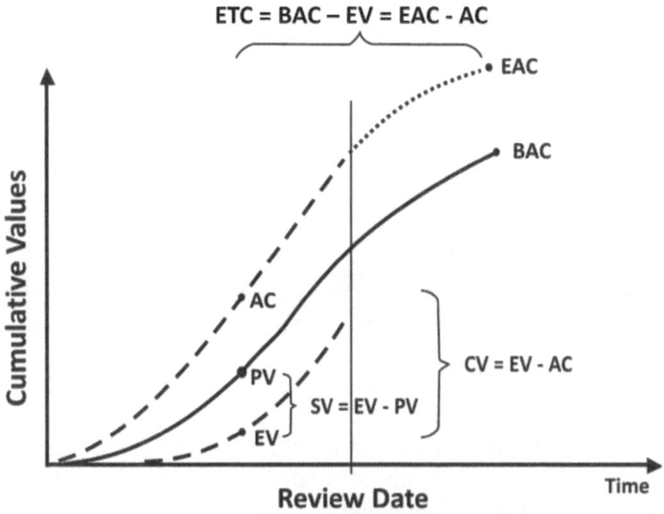

FIGURE 10.4 Earned Value Technique for Quality Assessment.

Examples of product factors are market share, flexibility, user acceptance, capacity utilization, safety, and design functionality. Factors are themselves considered to be composed of indicators. Examples of indicators are debt ratio, acquisition volume, product responsiveness, substitutability, lead time, learning curve, and scrap volume. By combining the above definitions, a composite measure of the operational value of a product can be quantitatively assessed. In addition to the quantifiable factors, attributes, and indicators that impinge upon overall project value, the human-based subtle factors should also be included in assessing overall project value.

EARNED VALUE TECHNIQUE FOR EARNED QUALITY

Value is synonymous with quality. Thus, the contemporary earned value technique is relevant for "earned quality" analysis. This is a good analytical technique to use for the justification stage of the DEJI model. This will impact cost, quality, and schedule elements of product development with respect to value creation. The technique involves developing important diagnostic values for each scheduled activity, work package, or control element as shown in Figure 10.4. The variables in the figure are PV: planned value; EV: earned value; AC: actual cost; CV: cost variance; SV: schedule variance; EAC: estimate at completion; BAC: budget at completion; ETC: estimate to complete. This analogical relationship is a variable research topic for quality engineering and technology applications.

INTEGRATION OF TECHNOLOGY

Without being integrated, a system will be isolated and it may be worthless. We must integrate all the elements of a system based on alignment of functional goals. The overlap of systems for integration purposes can conceptually be viewed as projection integrals by considering areas bounded by the common elements of sub-systems

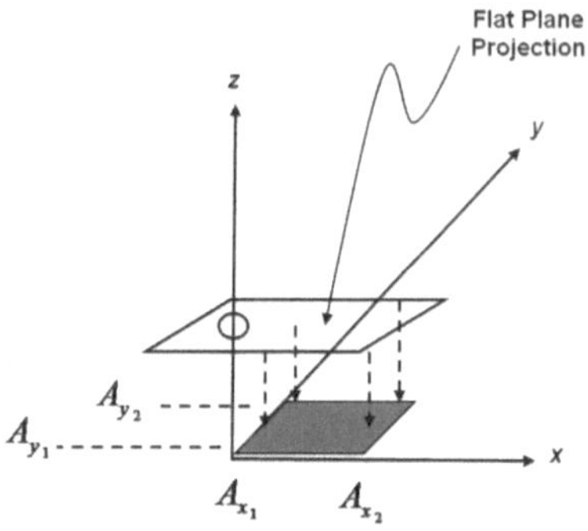

FIGURE 10.5 Technology Integration Surfaces.

as shown in Figure 10.5. Quantitative metrics can be applied at this stage for effective assessment of the product state. Trade-off analysis is essential in quality integration. Pertinent questions include the following:

What level of trade-offs on the level of quality are tolerable?
What is the incremental cost of higher quality?
What is the marginal value of higher quality?
What is the adverse impact of a decrease in quality?
What is the integration of quality of time? In this respect, an integral of the form below may be suitable for further research:

$$I = \int_{t_1}^{t_2} f(q)\,dq,$$

where I = integrated value of quality, $f(q)$ = functional definition of quality, t_1 = initial time, and t_2 = final time within the planning horizon.

Presented below are guidelines and important questions relevant to quality integration.

- What are the unique characteristics of each component in the integrated system?
- How do the characteristics complement one another?
- What physical interfaces exist among the components?
- What data/information interfaces exist among the components?
- What ideological differences exist among the components?
- What are the data flow requirements for the components?
- What internal and external factors are expected to influence the integrated system?
- What are the relative priorities assigned to each component of the integrated system?
- What are the strengths and weaknesses of the integrated system?
- What resources are needed to keep the integrated system operating satisfactorily?
- Which organizational unit has primary responsibility for the integrated system?

The systems approach of the DEJI model will facilitate a better alignment of product technology with future development and needs. The stages of the model require research for each new product with respect to design, evaluation, justification, and integration. Existing analytical tools and techniques as well as other systems engineering models can be used at each stage of the model. Thus, a hybrid systems modeling is possible. Quality is an integrative process that must be evaluated on a stage-by-stage approach. This requires research, education, and implementation strategies that consider several pertinent factors. This column suggests the DEJI model, which has been used successfully for product development applications, as a viable methodology for quality design, quality evaluation, quality justification, and quality integration. This column is intended as a source to spark the interest of researchers to apply this tool in new product development efforts.

APPLICATION GUIDE FOR SDGs AND GRAND CHALLENGES

A template for applying the DEJI systems model to UN's SDGs and NAE's grand challenges can be summed up in Figure 10.6. The dimensions of applications of the DEJI systems model cover the following elements:

- Productivity
- Efficiency
- Effectiveness
- Integration

The interweaving ovals in the figure signify non-discrete relationships among the design, evaluation, justification, and integration elements of the model (Badiru, 2013, 2019). The user will determine what factors or indicators fall within each stage of implementing the model and how those factors interrelate with other factors in the other stages.

In the context of the model, design implies the specification of requirements. Without well-designed, well-defined, and well-articulated requirements, all subsequent

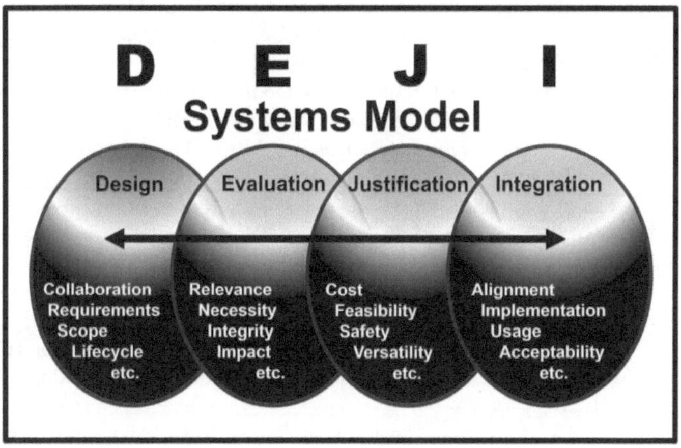

FIGURE 10.6 DEJI Systems Model Application Ovals.

efforts will be misguided. The modular components represent plug-and-play properties, implying systems interchangeability. DEJI systems model facilitates system modularity. Items designed in modules tend to offer more flexibility and adaptability. This is why industrial engineering is a versatile approach to system implementations of UN's SDGs and NAE's grand challenges. With the structured framework of DEJI systems model, technology implementations can be executed faster, more efficiently, effectively, and economically. Integration is the hallmark of the model. Integration should be done within the local context (alignment with prevailing scenarios) to maximize the possibility of success. Organizations are continuing to insert this highly functional framework into nearly every aspect of modern technology management. Everything is connected and very interdependent now, so everything now requires a systems engineering approach. DEJI is a flagship framework for planning contemporary projects, systems, teams, and technology ventures. Many organizations are excited about a lot of things related to new technology, but they don't have the foggiest idea of how to implement and integrate technology into their existing operations seamlessly. For example, safety by design is enabled by the model's structural methodology.

Design Quote:

Design is not how it looks and feels, it's how it works:

– Steve Jobs

"How it works" is a function of how it integrates with other work elements within the organization. In this sense, the design or specifications of the UN SDGs and NAE's grand challenges must align with the prevailing situations in each target area.

REFERENCES

Badiru, Adedeji B. (2012), "Application of the DEJI Model for Aerospace Product Integration," *Journal of Aviation and Aerospace Perspectives (JAAP)*, Vol. 2, No. 2, pp. 20–34, Fall 2012.

Badiru, Adedeji B. (2013), "Up to the Task: Industrial engineering is essential to meeting NAE's grand challenges," *Industrial Engineer*, Vol. 45, Issue 11, pp. 42–45, Nov 2013.

Badiru, Adedeji B. (2019), *Systems Engineering Models: Theory, Methods, and Applications*, Taylor & Francis Group/CRC Press, Boca Raton, FL.

Bhargava, V. editor, (2006), *Global Issues for Global Citizens: An Introduction to Key Development Challenges*, The World Bank, Washington, DC.

Bryant, Tamera (2005). *The Life & Times of Hammurabi*, Mitchell Lane Publishers, Newark, DE.

Duckworth, H. A. and R. A. Moore (2010), *Social Responsibility: Failure Mode Effects and Analysis*, Taylor and Francis Group/CRC Press, Boca Raton, FL.

Grayson, L. P. (1993), *The Making of an Engineer: An Illustrated History of Engineering Education in the United States and Canada*, John Wiley, New York, NY.

Kirby, R. S., S. Withington, A. B. Darlin, and F. G. Kilgour (1990), *Engineering in History*, Dover Publications, New York, NY.

Miller, Richard K. (2010), "The Educational Imperatives of the Engineering Grand Challenges, "Keynote Address, ASEE Global Colloquium on Engineering Education", Singapore, October 19–21, 2010.

Xian, H. and K. Madhavan (2014), "Anatomy of Scholarly Collaboration in Engineering Education: A Big-Data Bibliometric Analysis," *Journal of Engineering Education*, Vol. 103, No. 3, pp. 486–514.

11 Back to the Case Study of Space Shuttle Challenger

Technology Decision Disaster

CASE STUDY INTRODUCTION

This case study illustrates some of the problems involved in managing high-technology projects. The case study, from Badiru (1996), involves the events leading to the Space Shuttle Challenger accident that occurred on January 28, 1986. Although the accident happened several years ago, it continues to serve as an excellent example of a high-tech project failure. Its repercussions are still being felt in the space programs around the world. Readers should pay attention to the problems in the planning, conceptualization, technology assessment, communication, cooperation, and coordination processes of the Challenger mission. Figure 11.1 shows the Space Shuttle Challenger, in its initial regal pose on the launch pad.

BACKGROUND

The case study is based on the report of the Presidential Commission that investigated the Space Shuttle Challenger accident. The commission was chaired by William P. Rogers. The report was published in June, 1986, and it is available to the public as a matter of public record. Having the shuttle accident as a case study in a book helps to increase public awareness, and provides a formal setting for a critical analysis of a complex management process. The full Presidential Commission report (in V volumes) is overwhelming with pieces of managerial details logged in various sections. Thus, a case study provides a unified and condensed presentation of the management aspects of the accident. The case study can serve as a format for the managerial analysis of a high-tech system.

On September 30, 1986, Mr. B. I. Edelson, Associate Administrator with Space Science and Applications, wrote an open letter to the science community, requesting thoughts and advice on problems of the space program in the wake of the Challenger accident. This case study is a form of response to that request. The author hopes readers can develop good managerial analyses (or advice) from the case study to form a

DOI: 10.1201/9781032659220-11

FIGURE 11.1 Space Shuttle Challenger on the Launch Pad.

guideline for enhancing future operations not only for the space program but also for other high-tech and manufacturing endeavors.

Most of the materials in the case study are direct excerpts from the executive summary of the Presidential Commission report. The testimonies, in particular, are reproduced as published (verbatim) in the report to preserve the integrity of the original commission hearing. The quoted testimonies are indented and printed in small print to distinguish them from other text.

Points for discussion or reader analysis are printed in italics at the appropriate places in the case study. It is recommended that readers participate in the discussion process so as to develop competence in analyzing managerial decision problems. The scenarios of shuttle management should serve as models for readers to use in evaluating problems in their own operating environments. Efforts have been made to include as much of the testimony as possible to allow readers to have enough information to conduct the recommended managerial analysis.

THE SPACE SHUTTLE CONCEPT

The space shuttle concept originated in the 1960s during the development of the Apollo lunar landing spacecraft. The objective was to achieve economical access to space by using a reusable launch system. In September, 1969, a space task group offered a choice of three long-range plans:

1. A program costing $80 to $10 billion per year involving a manned Mars expedition, a space station in lunar orbit and a 50-person Earth-orbiting station serviced by a reusable ferry, or space shuttle.
2. An intermediate program that would include the Mars mission (less than $8 billion annually).
3. A modest program ($4 to $5.7 billion a year) that would embrace an Earth-orbiting space station with the space shuttle as its link to Earth.

In March, 1970, President Nixon opted for the shuttle-serviced space station as a long-range goal pending the development of the shuttle vehicle. In effect, the shuttle was, originally, simply the transportation element in a broad, multi-objective space plan. To proceed toward the ultimate plan, the shuttle would have to be built. Thus, the space shuttle became the major short-term focus of NASA; a focus that is really a means of another end.

DESIGN DECISIONS

The space shuttle went through evolutionary design changes. The first design was a "fly back" concept in which two manned stages are integrated. The first stage was a big rocket-powered vehicle that would carry the smaller stage piggyback. The carrier would provide the thrust for lift-off and flight through the atmosphere. It would release the passenger (the orbiting vehicle) and return to Earth. The orbiter, with its crew and payload, would continue into space under its rocket power. After completing its mission, it would fly back to Earth. The two-stage design carried its rocket propellants internally and had a much larger flight deck, and cargo bay than the later designs. The two-stage craft seemed to be an efficient means of achieving routine economical flight to space. However, the size of the craft called for huge development costs ($10 to $13 billion). Thus, it did not receive support in both congress and office of management and budget.

In 1971, NASA, now aware that low cost rather than system capability would garner support for the shuttle plan, went back to the drawing board. One proposal that emerged was to eliminate the internal tanks and carry the propellant in a single, disposable external tank. This allowed for a smaller and cheaper orbiter without significant performance loss. For the launch system, one proposal was a winged but unmanned recoverable liquid-fuel vehicle based on the successful Saturn-5 rocket from the Apollo program. Other plans considered simpler but also recoverable liquid-fuel systems, expendable solid rockets, and reusable solid rocket boosters. Solid-fuel systems had been used in the past for some small unmanned spacecraft, but using them for manned flight was a technology new to the space program. However, the solid rocket booster won approval over the liquid rocket which offered lower operating costs, but a higher development cost. The present space shuttle configuration was what emerged from this round of design efforts. The space shuttle is a three-component system made up of:

1. The orbiter.
2. An expendable external fuel tank carrying liquid propellants for the orbiter's engines.
3. Two recoverable solid rocket boosters.

A five-orbiter space shuttle system was estimated to have a cost of $6.2 billion in 1972, to develop and test. This was about half the cost of the two-stage "fly back" design. To achieve this lower development cost, NASA had to accept higher operating costs and sacrifice full reusability. The compromise design retained the recoverability and reuse of two of the three elements. The final configuration was selected in March 1972.

THE DEVELOPMENT PROCESS

In August, 1972, Rockwell International Corporation was awarded a contract for the design and development of the space shuttle orbiter. Martin Marietta was assigned the development and fabrication of the external tank, Morton Thiokol Corporation was awarded the contract for the solid rocket boosters, and Rocketdyne, a division of Rockwell, was selected to develop the orbiter main engines.

Managerial responsibility for the program was divided among three of NASA's field centers. Johnson Space Center in Houston was assigned the management of the orbiter. Marshall Space Flight Center in Huntsville, Alabama was responsible for the orbiter's main engines, Kennedy Space Center in Florida had the job of assembling the space shuttle components, checking them out, and conducting launches.

Discussion:

1. What coordination or decision problems may arise from the three-pronged managerial setup described above?
2. What could the separate contractors of the shuttle system do to ensure a unified design of the overall system?

The shuttle development years of the 1970s faced budgetary difficulties. The envisioned five-orbiter fleet was reduced to four. These difficulties were compounded by engineering problems. System changes that pushed the frontiers of technology led to cost overruns and schedule slippages. The initial test flights were delayed by more than two years. The first test flights were finally conducted at Dryden Flight Research Facility, California in 1977. The orbiter enterprise was the test craft. Carried piggyback on a modified Boeing 747, the enterprise was carried up and released for a gliding approach and landing at the Mojave Desert test center. Five such flights were made. They served to validate the computer operations, subsonic handling, and unpowered landing performance. Extensive ground tests were conducted from 1977 to 1980. These included system vibration tests and main engine firings.

By early 1981, the space shuttle was ready for an orbital flight test program. The test flights covered over 1000 tests and data collection procedures. All flights were to be launched from Kennedy and terminate at Edwards Air Force Base. The test flight series originally called for six missions. This was later reduced to four as shown below:

1. April 12–14, 1981: STS-1 (Space Transportation System-1). Orbiter Columbia went on a two-day demonstration of ability to go into orbit and return.
2. November 12–14, 1981: STS-2. Columbia tested the remote manipulator system and carried a payload of Earth survey instruments. The failure of a fuel cell shortened the flight by about three days.
3. March 22–30, 1982: STS-3. Columbia went on the longest of the test series. Special test of the robot arm was conducted and experiments in materials processing were performed.
4. June 27 to July 4, 1982: STS-4. Columbia carried the first Department of Defense payload. The conclusion of this flight ended the test flight program.

Ninety-five percent of the test objectives were accomplished. The time between flights was reduced from seven months to four, and then three. NASA then declared the space shuttle to be "operational," which meant that payload requirements in subsequent flights would be of more interest than spacecraft testing. The operational phase of the Space Shuttle Program, thus, began in November, 1982. The STS-sequential flight numbering was changed after STA-9 to two numbers followed by a letter; for example, 41-B. The first digit indicates the fiscal year of the scheduled launch (4 for 1984). The second digit identifies the launch site (1 for Kennedy, 2 for Vandenberg Air Force Base). The letter corresponds to the alphabetical sequence for the fiscal year. Thus, B means the second scheduled mission in the fiscal year. Due to schedule changes, some flights were not made according to the sequence of numbering. Thus, 51-D occurred before 51-B. Presented below is the chronology of shuttle flights after the test phase.

1. November 11–16, 1982: STS-5 (Columbia).
2. April 4–9, 1983: STS-6 (Challenger).
3. June 18–24, 1983: STS-7 (Challenger).
4. August 30 to September 6, 1983: STS-8 (Challenger).
5. November 28 to December 8, 1983: STS-9 (Columbia).
6. February 3–11, 1984: 41-B (Challenger).
7. April 6–13, 1984: 41-C (Challenger).
8. August 30 to September 5, 1984: 41-D (Discovery).
9. October 5–13, 1984: 41-G (Challenger).
10. November 8–16, 1984: 51-A (Discovery).
11. January 24–27, 1985: 51-C (Discovery).
12. April 12–19, 1985: 51-D (Discovery).
13. April 29 to May 6, 1985: 51-B (Challenger).
14. June 17–24, 1985: 51-G (Discovery).

15. July 29 to August 6, 1985: 51-F (Challenger).
16. August 27 to September 3, 1985: 51-I (Discovery).
17. October 3–10, 1985: 51-J (Atlantis).
18. October 30 to November 6, 1985: 61-A (Challenger).
19. November 26 to December 3, 1985: 61-B (Atlantis).
20. January 12–18, 1986: 61-C (Columbia).
21. January 28, 1986: 51-L (Challenger: The Accident).

The Space Shuttle Program completed 24 successful missions (including the four test flights) over 57-month period. Columbia made seven trips, Discovery made six, Atlantis two, and Challenger made nine trips prior to the accident. A shuttle crew may consist of up to eight people; but the limit is generally seven. The crew consists of the commander, the captain, the pilot, second in command, and two or more mission specialists. One or more payload specialists can be accommodated. A mission specialist coordinates activities of the craft and crew. A payload specialist may manage specific experiments. The commander, pilot, and mission specialists are career astronauts assigned to the mission by NASA. Payload specialists are assigned by payload sponsors in coordination with NASA.

THE CHALLENGER MISSION

Changes in the Challenger launch schedule complicated the preparations for mission 51-L. The sequence of interrelated activities involved in producing the detailed schedule, and supporting logistics necessary for a successful mission requires considerable level of effort and close coordination. Flight 51-L was originally scheduled for July, 1985, but schedule slippages forced the delay till January, 1986. Planning for the mission began in 1984. Ten major changes in payload items caused disruption in the preparation process. Because the 12 to 18 months of preparation involve repetitive cycles that progressively define a flight plan in more specific detail, significant changes can require extensive time and effort to incorporate. The closer to the launch the changes occur, the more difficult and disruptive it becomes to repeat the preparation cycles.

LAUNCH DELAYS

The launching of 51-L was postponed three times and scrubbed once from the planned date of January 22, 1986. The launch first slipped from January 23 to January 25. That date was later changed to January 26, 1986, primarily because of Kennedy Space Center work requirements caused by the delay in the launching of mission 61-C (Columbia). The third postponement of the launch date occurred on the evening of January 25 due to a forecast of bad weather for January 26. The launch was rescheduled for January 27. At 9:10 a.m. on the new launch day, the countdown was halted, when the ground crew reported a problem with an exterior hatch handle. By the time the problem was solved at 10:30 a.m., winds at the runway designated for a "return-to-launch-site" abort had picked up considerably and exceeded the allowable crosswinds. The launch attempt for January 27, was thus canceled at 12:35 p.m. of that day. The countdown was rescheduled for January 28.

Discussion: Based on the long record of launch delays, should NASA have temporarily suspended all launches in order to conduct extensive evaluations of the technical or managerial problems? Present arguments to support both the pros and cons of the decision to continue launching.

LAUNCH PROBLEMS

The weather for January 28 was forecast to be very cold, with overnight temperatures in the low twenties. The management team directed engineers to assess the possible effects of temperature on the launch. No critical issues were identified to the management officials. So, countdown proceeded while weather evaluation continued. Ice had accumulated in the launch pad area during the night, and it caused considerable concern for the launch team. The ice inspection team was sent to the pad at 1:35 a.m., January 28, and returned to the launch control center at 3:00 a.m. After reviewing the team's report at a program meeting, the Space Shuttle Program Manager decided to continue the countdown. Another ice inspection was scheduled at launch -3 hours. During the night, prior to fueling, a problem developed with a fire detector in the ground liquid hydrogen storage tank. Though the problem was tracked to a hardware fault and repaired, fueling was delayed by two and one-half hours. However, the launch delay was reduced to only one hour by continuing past, a planned hold at launch -3 hours. Because of rain forecast at Casablanca, the alternate abort site, that site was scrubbed at 7:30 a.m. The change had no impact on the mission since the weather at the primary transatlantic abort landing site at Dakar, Senegal, was acceptable.

At a flight crew weather briefing early in the morning, the temperature and ice on the pad were discussed. But neither then, nor in earlier weather discussions was the crew told of any concern about the effect of low temperature on the shuttle system. By 8:36 a.m., all the seven crew members (Francis R. Scobee, Michael J. Smith, Ellison S. Onizuka, Judith A. Resnik, Ronald E. McNair, Christa McAuliffe, and Gregory B. Jarvis) were in their seats in the Challenger. At 8:44 a.m., the ice team completed its second inspection. After hearing their report, the program manager decided to allow additional time for ice to melt on the pad. He also told the ice team to perform one final ice assessment at launch -20 minutes. By the time the countdown resumed, the launch had been delayed a second hour beyond the lift-off time of 9:38 a.m. The final ice inspection was completed at 11:15 a.m. During the hold at launch -9 minutes, the mission crew and all members of the launch team gave their approval for launch. The flight began at 11:38 a.m., January 28, 1986.

Discussion: Should the crew have had a say on whether or not to launch?

THE ACCIDENT

There was no indication of problem from lift-off until the signal from the shuttle was lost. The main engines operated satisfactorily as expected. Voice communications with the crew were normal. The crew called to indicate the shuttle had begun its roll to head due east and to establish communication after launch. Fifty-seven seconds later, mission control informed the crew that the engines had successfully throttled up and all other systems were satisfactory. The commander's acknowledgment of this call was the last voice communication from the Challenger.

No alarms were sounded in the cockpit. The crew apparently had no indication of a problem before the rapid explosion of the shuttle which occurred 73 seconds after lift-off. The first evidence of the accident came from live video coverage. There were no survivable abort options during the thrusting of the solid rocket boosters. There was nothing that either the crew or the ground controllers could have done to stop the catastrophe.

The consensus of the commission and other agencies that investigated the accident is that the accident was caused by a failure in the joint between the two lower segments of the right solid rocket motor (SRM). The specific failure was the destruction of the seals that are intended to prevent hot gases from leaking through the joint during the propellant burn of the rocket motor. There is no evidence that any other element of the shuttle contributed to the failure.

Discussion: Do you feel that a lack of coordinated hardware design contributed to the failure of the seals?

THE ACCIDENT INVESTIGATION

Many aspects of the Challenger mission were investigated by the Presidential Commission. All the technical and management factors were examined in detail. For this case study, only the management issues are addressed. The following presentations including testimonies are excerpts from the executive summary of the Presidential Commission. For a more detailed coverage, refer to the executive summary itself or the full commission report, volumes I to V.

FLAWS IN THE DECISION PROCESS

The decision to launch the Challenger was flawed. Those who made that decision, were unaware of the recent history of problems, concerning the O-rings and the joint, and were unaware of the initial written recommendation of the contractor advising, against the launch at temperatures below 53 degrees Fahrenheit, and the continuing opposition of the engineers at Thiokol, after the management reversed its position. They did not have a clear understanding of Rockwell's concern that it was not safe to launch because of ice on the pad. If the decision-makers had known all of the facts, it is highly unlikely that they would have decided to launch 51-L on January 28, 1986. The discussion that follows is based on excerpts from the testimony, of those involved in the management judgments, that led to the launch of the Challenger under the conditions described. That testimony reveals failures in communication that resulted in a decision to launch 51-L, based on incomplete and sometimes misleading information, a conflict between engineering data and management judgments, and a NASA management structure that permitted internal flight safety problems to bypass key Shuttle managers.

The Shuttle Flight Readiness Review is a step-by-step activity, established by NASA to certify the readiness of all components of the Space Shuttle assembly. The process is focused on the Level I Flight Readiness Review, held approximately two weeks before a launch. The Level I review (see Figure 11.2) is a conference chaired

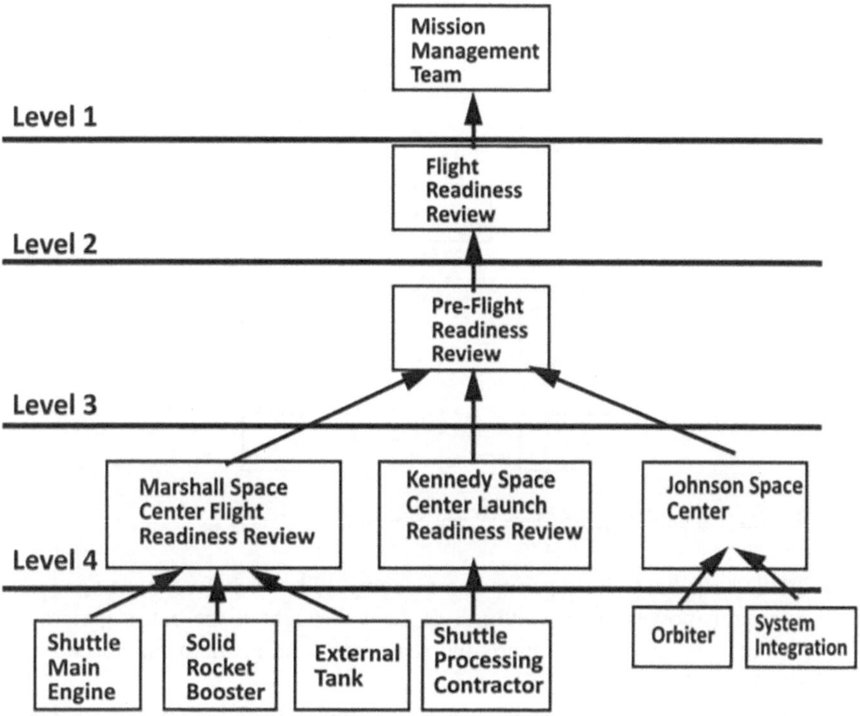

FIGURE 11.2 Flight Readiness Review Hierarchies.

by the NASA Associate Administrator for Space Flight and supported by the NASA Chief Engineer, the Program Manager, the center directors, and project managers from Johnson, Marshall, and Kennedy, along with senior contractor representatives.

The formal portion of the process is initiated by directive from the Associate Administrator for Space Flight. The directive outlines the schedule for the Level I Flight Readiness Review and for the steps that precede it. The process begins at Level IV (see Figure 11.3) with the contractors formally certifying (in writing) the flight readiness of the elements for which they are responsible. Certification is made to the appropriate Level III NASA project managers at Johnson and Marshall. Additionally, at Marshall, the review is followed by a presentation directly to the Center Director. At Kennedy, the Level III review, chaired by the Center Director, verifies readiness of the launch support elements.

The next step in the process is the Certification of Flight Readiness to the Level II Program Manager at Johnson. In this review, each Space Shuttle Program element endorses that it has satisfactorily completed the manufacture, assembly, test, and checkout of the pertinent element, including the contractors' certification that design and performance are up to standard. The Flight Readiness Review process culminates in the Level I review.

In the initial notice of the review, the Level I directive establishes a Mission Management Team for the particular mission. The team assumes responsibility for

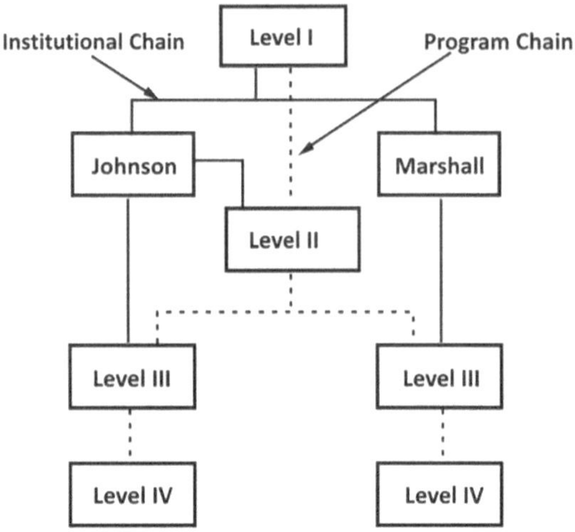

FIGURE 11.3 Shuttle Program Management Structure.

each Shuttle's readiness for a period commencing 48 hours before launch and continuing through post-landing activities. On call throughout the entire period, the Mission Management Team supports the Associate Administrator for Space Flight and the Program Manager.

A structured Mission Management Team meeting, called L-1, is held 24 hours, before each scheduled launch. Its agenda includes a closeout of any open work, a closeout of any Flight Readiness Review action items, a discussion of new or continuing anomalies, and an updated briefing on anticipated weather conditions at the launch site, and at the abort landing sites in different parts of the world. It is standard practice of Level I and II officials, to encourage the reporting of new problems or concerns that might develop in the interval between the Flight Readiness Review and the L-1 meeting, and between the L-1 launch. In a procedural sense, the process described was followed in the case of flight 51-L. However, in the launch preparation for 51-L, the relevant concerns of Level III NASA personnel and element contractors were not adequately communicated to the NASA Level I and II management responsible for the launch.

Discussion: It seems NASA has a well-established and potentially effective managerial system. However, the implementation of the system might have been flawed. Why did that standard system not work for Challenger? Many things that could go wrong did just that. Is Murphy's Law manifesting itself here?

Two of the specific concerns that were not properly addressed in the Challenger preparations are:

1. The objections to launch voiced by Morton Thiokol engineers, about the detrimental effect of cold temperatures on the performance of the SRM joint seal.
2. The degree of concern of Thiokol and Marshall, about the erosion of the joint seals in prior Shuttle flights, notably 51-C (January, 1985) and 51-B (April, 1985).

On January 23, Jesse Moore, associate administrator for space flight, issued a directive stating that the Flight Readiness Review had been conducted on January 15 and that 51-L was ready to fly pending closeout of open work, satisfactory countdown, and completion of remaining Flight Readiness Review action items, which were to be closed out during the L-1 meeting. No problems with the Solid Rocket Booster were identified.

Since December, 1982, the O-rings had been designated a "Criticality 1" feature of the Solid Rocket Booster design, a term denoting a failure point (without back-up) that could cause a loss of life or vehicle if the component fails. In July, 1985, after a nozzle joint on STS 51-B showed erosion of a secondary O-ring, indicating that the primary seal failed, a launch constraint was placed on flight 51-F and subsequent launches. These constraints had been imposed and regularly waived by the Solid Rocket Booster Project Manager at Marshall, Lawrence B. Mulloy. Neither the launch constraint, the reason for it, nor the six consecutive waivers prior to 51-L were known to Moore (Level I), Aldrich (Level II), or Thomas at the time of the Flight Readiness Review process for 51-L.

There were other paths of system reporting that were designed to bring forward information about the Solid Rocket Booster joint anomalies. One path was the task force of Thiokol engineers and Marshall engineers, who had been conducting subscale pressure tests during 1985; a source of documented rising concern and frustration on the part of some of the Thiokol participants and a few of the Marshall participants. However, Level II was not in the line of reporting for this activity. Another path was the examination at each Flight Readiness Review of evidence of earlier flight anomalies. For 51-L, the data presented in this latter path, while it reached Levels I and II, never referred to either test anomalies or flight anomalies with O-rings.

No mention of the O-ring problems in the Solid Rocket Booster joint, appeared in the Certification of Flight Readiness, signed for Thiokol on January 9, 1986, by Joseph Kilminster, for the Solid Rocket Booster set designated BI026. Similarly, no mention appeared in the certification endorsement, signed on January 15, 1986, by Kilminster and Mulloy. In the 51-L readiness reviews, it appears that neither Thiokol management nor the Marshall Level III project managers believed that the O-ring blow-by and erosion risk was critical. The testimony and contemporary correspondence show that Level III believed there was ample margin to fly with O-ring erosion, provided the leak check was performed at 200 pounds per square inch. Following the January 15, Flight Readiness Review each element of the Shuttle was certified as flight-ready. The Level I Mission Management Team meeting took place as scheduled at 11:00 a.m. on January 25. No technical issues appeared at this meeting or in the documentation, and all Flight Readiness Review actions were reported closed out.

Discussion:

1. Do the several layers of managerial reporting impede critical communication requirements? If so, what is a possible remedy?
2. How could the seriousness of the ring problem have been better impressed upon the Thiokol management and the Marshall project managers?

THE TESTIMONIES

Mr. Mulloy testified as follows regarding the Flight Readiness Review record about O-ring concerns:

Chairman Rogers: ... Why wasn't that a cause for concern on the part of the whole NASA organization?
Mr. Mulloy: It was cause for concern, sir.
Chairman Rogers: Who did you tell about this?
Mr. Mulloy: Everyone, sir.
Chairman Rogers: And they all knew about it at the time of 51-L?
Mr. Mulloy: Yes, sir. You will find in the Flight Readiness Review record that went all the way to the L-1 review.

But contrary to the testimony of the Solid Rocket Booster Project Manager, the seriousness of concern was not conveyed in Flight Readiness Review to the Level I personnel. The only remaining issue facing the Mission Management Team at the L-1 review was the approaching cold front, with forecasts of rain showers and temperatures in the mid-sixties. There had also been heavy rain since 51-L had been rolled out to the launch pad, approximately seven inches compared with the 2.5 inches that would have been normal for that season. At 12:36 p.m. on the 27th, the Mission Management Team scrubbed the launch for that day due to high crosswinds at the launch site. In the accompanying discussion that ran for about half an hour, all appropriate personnel were polled as to the feasibility of a launch within 24 hours. Participants were requested to identify any constraints. This meeting, aimed at launch at 9:38 a.m. on January 28, produced no constraints or concerns about the performance of the Solid Rocket Boosters.

At 2:00 p.m. on the 27th, the Mission Management Team met again. At that time, the weather was expected to clear, but it appeared that temperatures would be in the low twenties for about 11 hours. Issues were raised about the cold weather effects on the launch facility, including the water drains, eyewash, and shower water, the fire suppression system, and overpressure water trays. It was decided to activate heaters in the Orbiter, but no concerns were expressed about the O-rings in the Solid Rocket Boosters. The decision was to proceed with the countdown and with fueling, but all members of the team were asked to review the situation and call if any problem arose.

At about 2:30 p.m., at Thiokol's Wasatch plant, Robert Ebeling, after learning of the predicted low temperature for launch, convened a meeting with Roger Boisjoly and with other Thiokol engineers. Ebeling was concerned about predicted cold temperatures at Kennedy Space Center. In a post-accident interview, Mr. Ebeling recalled the substance of the meeting.

> The meeting lasted one hour, but the conclusion of that meeting was Engineering—especially Arnie, Roger Boisjoly, Brian Russell, myself, Jerry Burns, they come to mind—were very adamant about their concerns on this lower temperature, because we were way below our data base, and we were way below what we qualified for.

Later in the afternoon on the same day, Allan McDonald—Thiokol's liaison for the Solid Rocket Booster project at Kennedy Space Center—received a telephone call from Ebeling, expressing concern about the performance of the Solid Rocket Booster field joints at low temperatures. During testimony before the Commission on February 27, McDonald recounted that conversation:

Mr. McDonald: Well, I had first become aware of the concern of the low temperatures that were projected for the Cape, it was late in the afternoon on the 27th. I was at Carver Kennedy's house. He is a vice president of, as I mentioned, our space operations center at the Cape, and supports the stacking of the solid rocket motors (SRMs). And I had a call from Bob Ebeling. He is the manager of our ignition system and final assembly, and he worked for me as a program manager at Thiokol in Utah. He called me and said that they had just received some word earlier that the weatherman was projecting temperatures as low as 18 degrees Fahrenheit some time in the early morning hours of the 28th, and that they had some meetings with some of the engineering people and had some concerns about the O-rings getting to those kinds of temperatures.

And he wanted to make me aware of that, and also wanted to get some more updated and better information, on what the actual temperature was going to be depicted so that they could make some calculations on what they expected the real temperature the O-rings may see…

I told him that I would get that temperature data for him and call him back. Carver Kennedy then, when I hung up, called the launch operations center to get the predicted temperatures from pad B, as well as what the temperature history had been during the day up until that time.

… He obtained those temperatures from the launch operations center, and they basically said that they felt it was going to get near freezing, or freezing before midnight. It would get as low as 22 degrees as a minimum in the early morning hours, probably around 6:00 o'clock, and they were predicting a temperature of about 26 degrees at the intended time, about 9:38 the next morning.

I took that data and called back to the plant and sent it to Bob Ebeling and relayed that to him, and told him he ought to use this temperature data for his predictions, but I thought this was very serious and to make sure that he had the vice president, engineering, involved in this and all of his people; that I wanted them to put together some calculations and a presentation of material.

Chairman Rogers: Who's the Vice President, Engineering?

Mr. McDonald: Mr. Bob Lund is our Vice President, Engineering, at our Morton Thiokol facility in Utah.

To make sure he was involved in this, and that this decision should be an engineering decision, not a program management decision. I told him that I would like him to make sure they prepared some charts and were in a position to recommend the launch temperature and to have the rationale for supporting that launch temperature.

I then hung up and called Mr. Mulloy. He was staying at the Holiday Inn in Merritt Island and they couldn't reach him, so I called Cecil Houston—Cecil Houston is the resident manager for the Marshall Space Flight Center office at Kennedy Space Center (KSC)—and told him about our concerns with the low temperatures and the potential problem with the O-rings.

And he said that he would set up a teleconference. He had a four-wire system next to his office. His office is right across from the VAB (vehicle assembly building) in the trailer complex C over there. And he would set up a four-wire teleconference involving the engineering people at Marshall Space Flight Center at Huntsville, our people back at Thiokol in Utah; and that I should come down to his office and participate at Kennedy from there, and that he would get back with me and let me know when that time would be.

Soon thereafter Cecil Houston called Dr. Judson Lovingood, Deputy Shuttle Project Manager at Marshall Space Flight Center, to inform him of the concerns about the O-rings and asked Lovingood to set up a teleconference with senior project management personnel, with George Hardy, Marshall's Deputy Director of Science and Engineering, and with Morton Thiokol personnel. Lovingood called Stanley Reinartz, Shuttle Project Manager, a few minutes later and informed him of the planned teleconference. The first phase of the teleconference began at 5:45 p.m. Eastern Standard Time; participants included Reinartz, Lovingood, Hardy, and numerous people at Kennedy, Marshall and Thiokol-Wasatch. (Allan McDonald missed this phase; he did not arrive at Kennedy until after 8:00 p.m.). Concerns for the effect of low temperature on the O-rings and the joint seal were presented by Morton Thiokol, along with an opinion that the launch should be delayed. A recommendation was also made that Aldrich, Program Manager at Johnson (Level II), be informed of these concerns. The following are excerpts from testimony before the Presidential Commission relating to the teleconference:

Dr. Keel: You just indicated earlier that, based upon that teleconference, you thought there was a good possibility of delay. Is that what Thiokol was recommending then, was delay?

Dr. Lovingood: That is the way I heard it, and they were talking about the 51-C experience and the fact that they had experienced the worst-case blow-by as far as the arc and the soot and so forth. And also, they talked about the resiliency data that they had. So it appeared to me--and we didn't have all of the proper people there. That was another aspect of this. It appeared to me that we had better sit down and get the data so that we could understand exactly what they were talking about and assess that data. And that is why I suggested that we go ahead and have a telecon within the center so that we could review that.

Dr. Keel: So as early as after that first afternoon conference at 5:45, it appeared that Thiokol was basically saying delay. Is that right?

Dr. Lovingood: That is the way it came across to me. I don't know how other people perceived it, but that's the way it came across to me.

Dr. Keel: Mr. Reinartz, how did you perceive it?

Mr. Reinartz: I did not perceive it that way. I perceived that they were raising some questions and issues which required looking into by all the right parties, but I did not perceive it as a recommendation delay.

Dr. Keel: Some prospects for delay?

Mr. Reinartz: Yes, sir, that possibility is always there.

Dr. Keel: Did you convey that to Mr. Mulloy and Mr. Hardy before the 8:15 conference?

Mr. Reinartz: Yes, I did. And as a matter of fact, we had a discussion. Mr. Mulloy was just out of communication for about an hour, and then after that, I got in contact with him, and we both had a short discussion relating to the general nature of the concerns with Dr. Lucas and Mr. Kingsbury at the motel before we both departed for the telecon that we had set up out at the Cape.

Dr. Keel: But based upon that, Mr. Lovingood, that impression, you thought it was a significant enough possibility that Mr. Aldrich should have been contacted?

Dr. Lovingood: Yes.

Dr. Keel: In addition, did you recommend that Mr. Lucas, who is the director of Marshall, of course, and Mr. Kingsbury who is Mr. Hardy's boss, participate in the 8:15 conference?

Dr. Lovingood: Yes, I did.

Dr. Keel: And you recommended that to whom?

Dr. Lovingood: I believe I said that over the net. I said that I thought we ought to have an inter-center meeting involving Dr. Lucas and Mr. Kingsbury, and then plan to go up the line to Level II and Level I.

And then it was after we broke off that first telecon I called Stan to the motel and told him that he ought to go ahead and alert Arnie to that possibility.

Dr. Keel: And Mr. Reinartz, you then visited the motel room of Mr. Lucas with Mr. Kingsbury, and also was Mr. Mulloy with you then?

Mr. Reinartz: Yes, sir, he was. In the first couple of minutes I believe I was there by myself, and then Mr. Mulloy joined us.

Dr. Keel: And did you discuss with them Mr. Lovingood's recommendation that the two of them, Lucas and Kingsbury, participate?

Mr. Reinartz: No, sir. I don't recall discussing Mr. Lovingood's recommendations. I discussed with them the nature of the telecon, the nature of the concerns raised by Thiokol, and the plans to gather the proper technical support people at Marshall for examination of the data. And I believe that was the essence of the discussion.

Chairman Rogers: But you didn't recommend that the information be given to Level II or Level I?

Mr. Reinartz: I don't recall that I raised that issue with Dr. Lucas. I told him what the plans were for proceeding. I don't recall, Mr. Chairman, making any statement regarding that.

Mr. Hotz: Mr. Reinartz, are you telling us that you in fact are the person who made the decision not to escalate this to a Level II item?

Mr. Reinartz: This is correct, sir.

Discussion: Is Mr. Reinartz's judgment appropriate? What should he have done?

At approximately 8:45 p.m., Phase 2 of the teleconference commenced, the Thiokol charts and written data having arrived at Kennedy Space Center by telefax. (A table of teleconference participants is included with the Chronology of Events). The charts presented a history of the O-ring erosion and blow-by in the Solid Rocket Booster joints of previous flights, presented the results of subscale testing at Thiokol and the results of static tests of SRMs. In the following testimony, Roger Boisjoly, Allan McDonald and Larry Mulloy expressed their recollections of this teleconference up to the point when an off-net caucus was requested:

Mr. Boisjoly: I expressed deep concern about launching at a low temperature. I presented Chart 2-1 with emphasis--now, 2-1, if you want to see it, I have it, but that was the chart that summarized the primary concerns, and that was the chart that I pulled right out of the Washington presentation without changing one word of it because it was still applicable, and it addresses the highest concern of the field joint in both the ignition transient condition and the steady state condition, and it sets down the rationale for why we were continuing to fly. Basically, if erosion penetrates the primary O-ring seal, there is a higher probability of no secondary seal capability in the steady state condition. I had two sub-bullets under that which stated bench testing showed the O-ring was not capable of maintaining contact with metal parts, gap, opening rate to maximum operating pressure. I had another bullet that stated bench testing showed the capability to maintain O-ring contact during the initial phase (O–170 milliseconds of transient). That was my comfort basis of continuing to fly under normal circumstances, normal being within the database we had.

I emphasized when I presented that chart about the changing of the timing function of the O-ring as it attempted to seal. I was concerned that we may go from that first beginning region into that intermediate region, from 0 to 170 being the first region, and 170–330 being the intermediate region where we didn't have a high probability of sealing or seating.

I then presented Chart 2–2 with added concerns related to the timing function. On that chart, I started off talking about a lower temperature than the current database results in changing the primary O-ring sealing timing function, and I discussed the SRM-15 (Flight 51-C, January, 1985) observations, namely, the 15A (Left SRM, Flight 51-C) motor had 80-degrees arc black grease between the O-rings, and make no mistake about it, when I say black, I mean black just like coal. It was jet black. SRM-15B (Right SRM, Flight 51-C) had a 110-degree arc of black grease between the 0-rings. We would have a low O-ring squeeze due to low temperature which I calculated earlier in the day. We should have higher O-ring Shore hardness…

Now, that would be harder. And what that material is, it would be likened to trying to shove a brick into a crack versus a sponge. That is a good analogy for this discussion. I also mentioned that thicker grease, as a result of lower temperatures, would have a higher viscosity. It wouldn't be as slick and slippery as it would be at room temperature. And so it would be a little bit more difficult to move across it.

We would have higher O-ring pressure actuation time, in my opinion, and that is what I presented….These are the sum and substance of what I just presented. If action time increases, then the threshold of secondary seal pressurization capability is approached. That was my fear. If the threshold is reached, then the secondary seal may not be capable of being pressurized, and that was the bottom line of everything that had been presented up to that point.

Chairman Rogers: Did anybody take issue with you?

Mr. Boisjoly: Well, I am coming to that. I also showed a chart of the joint with an exaggerated cross-section to show the seal lifted off, which has been shown to everybody. I was asked, yes, at that point in time I was asked to quantity my concerns, and I said I couldn't. I couldn't quantify it. I had no data to quantify it, but I did say I knew that it was away from goodness in the current data base. Someone on the net commented that we had soot blow-by on SRM-22 [flight 61-A, October, 1985] which was launched at 75 degrees. I don't remember who made the comment, but that is where the first comment came in about the disparity between my conclusion and the observed data because SRM-22 [Flight 61-A, October, 1985] had blown-by at essentially a room temperature launch.

I then said that SRM-15 [Flight 51-C, January, 1985] had much more blown-by indication and that it was indeed telling us that lower temperature was a factor. This was supported by inspection of flown hardware by myself. I was asked again for data to support my claim, and I said I have none other than what is being presented. I had been trying to get resilience data, Arnie and I both, since last October, and that statement was mentioned on the net.

Others in the room presented their charts, and the main telecon session concluded with Bob Lund, who is our Vice President of Engineering, presenting his conclusions and recommendations charts which were based on our data input up to that point. Listeners on the telecon were not pleased with the conclusions and the recommendations.

Chairman Rogers: What was the conclusion?

Mr. Boisjoly: The conclusion was we should not fly outside of our data base, which was 53 degrees. Those were the conclusions. And we were quite pleased because we knew in advance, having participated in the preparation, what the conclusions were, and we felt very comfortable with that.

Mr. Acheson: Who presented that conclusion?

Mr. Boisjoly: Mr. Bob Lund. He had prepared those charts. He had input from other people. He had actually physically prepared the charts. It was about that time that Mr. Hardy from Marshall was asked what he thought about the MTI (Morton Thiokol) recommendation, and he said he was appalled at the MTI decision. Mr. Hardy was also asked about launching, and he said no, not if the contractor recommended not launching, he would not go against the contractor and launch.

There was a short discussion that ensued about temperature not being a discriminator between SRM-15 [Flight 51-C] and SRM-22 [Flight 61-A], and shortly after, I believe it was Mr. Kilminster asked if--excuse me. I'm getting confused here. Mr. Kilminster was asked by NASA if he would launch, and he said no because the engineering recommendation was not to launch.

Then MTI management asked for a five-minute caucus. I'm not sure exactly who asked for that, but it was asked in such a manner that I remember it was asked for, a five-minute caucus, which we put on – [we put]the line on mute and went off-line with the rest of the net.

Chairman Rogers: Mr. Boisjoly, at the time that you made the--that Thiokol made the recommendation not to launch, was that the unanimous recommendation as far as you knew?

Mr. Boisjoly: Yes. I have to make something clear. I have been distressed by the things that have been appearing in the paper and things that have been said in general, and there was never one positive, pro-launch statement ever made by anybody. There have been some feelings since then that folks have expressed that they would support the decision, but there was not one positive statement for launch ever made in that room.

Mr. McDonald's Testimony

Mr. McDonald: I arrived at the Kennedy Space Center at about 8:15[p.m.], and when I arrived there at the Kennedy Space Center the others that had already arrived were Larry Mulloy, who was there--he is the manager, the project manager for the SRB for Marshall. Stan Reinartz was there and he is the manager of the Shuttle Project Office. He's Larry Mulloy's boss. Cecil Houston was there, the resident manager for Marshall. And Jack Buchanan was there. He happens to be our manager, Morton Thiokol's manager of our launch support services office at Kennedy. The telecon hadn't started yet. It came on the network shortly after I got there …

Chairman Rogers: Was it essentially a telephone conference or was there actually a network of pictures?

Mr. McDonald: It was a telephone conference ….

But I will relay…what I heard at the conference as best I can. The teleconference started I guess close to 9:00 o'clock and, even though all the charts weren't there, we were told to begin and that Morton Thiokol should take the lead and go through the charts that they had sent to both centers.

The charts were presented by the engineering people from Thiokol, in fact by the people that had made those particular charts. Some of them were typed, and some of them were handwritten. They discussed their concerns with the low temperatures relative to the possible effects on the O-rings, primarily the timing function to seal the O-rings.

They presented a history of some of the data that we had accumulated both in static tests and in flight tests relative to temperatures and the performance of the

O-rings, and reviewed the history of all of our erosion studies of the O-rings, in the field joints, any blown-by the primary O-ring with soot or products of combustion or decomposition that we had noted, and the performance of the secondary O-rings.

And there was an exchange among the technical people on that data as to what it meant…But the real exchange never really came until the conclusions and recommendations came in.

At that point in time, our vice president, Mr. Bob Lund, presented those charts and he presented the charts on the conclusions and recommendations. And the bottom line was that the engineering people would not recommend a launch below 53 degrees Fahrenheit. The basis for the recommendation was primarily our concern with the launch that had occurred about a year earlier, in January of 1985, I believe it was 51-C.

Mr. Mulloy's Testimony

Mr. Mulloy: That telecon was a little late starting. It was intended to be set up at 8:15…and the teleconference was begun at 8:45.

Thiokol will then present to you today that data that they presented to us in that telecon. I will not do that. The bottom line of that, though, initially was that Thiokol Engineering, Bob Lund, who is the Vice President and Director of Engineering, who is here today, recommended that 51-L not be launched if the O-ring temperatures predicted at launch time would be lower than any previous launch and that was 53 degrees.

Dr. Walker: May I ask a question? I wish you would distinguish between the predicted bulk temperatures and the O-ring temperatures. In fact, as I understand it, you really don't have any official O-ring temperature prediction in your models, and it seems that the assumption has been that the O-ring temperature is the same as the bulk temperature, which we know is not the case.

Mr. Mulloy: You will see, sir, in the Thiokol presentation today that that is not the case. This was a specific calculation of what the O-ring temperature was on the day of the January 1985 launch. It is not the bulk temperature of the propellant, nor is it the ambient temperature of the air.

It was Thiokol's calculation of the lowest temperature an O-ring had seen in previous flights, and the engineering recommendation was that we should not move outside of that experience base.

I asked Joe Kilminster, who is the program manager for the booster program at Thiokol, what his recommendation was, because he is the gentleman that I get my recommendations from in the program office. He stated that, based on that engineering recommendation, he could not recommend a launch.

At that point I restated, as I have testified to, the rationale that was essentially documented in the 1982 Critical Items List, which stated that the rationale had been that we were flying with a simplex joint seal. You will see in the Thiokol presentation that the context of their presentation is that the primary ring, with the reduced temperatures and reduced resiliency, may not function as a primary seal and we would be relying on a secondary.

And without getting into their rationale and getting ahead, the point, the bottom line, is that we were continuing--the assessment was, my assessment at that time was, that we would have an effective simplex seal, based upon the engineering data that Thiokol had presented, and that none of those engineering data seemed to change that basic rationale.

Stan Reinartz then asked George Hardy, the Deputy Director of Science and Engineering at Marshall, what his opinion was. George stated that he agreed that the engineering data did not seem to change this basic rationale, but also stated on the telecon that he certainly would not recommend launching if Thiokol did not.

At that time Joe Kilminster requested a five-minute off-net caucus, and that caucus lasted approximately 30 minutes.

The teleconference was recessed at approximately 10:30 p.m. The off-net caucus of Thiokol personnel started and continued for about 30 minutes at the Wasatch office. The major issues, according to the testimony of Jerry Mason, Senior Vice President for Wasatch Operations, were the effect of temperature upon the O-rings and the history of erosion of the O-rings:

Mr. Mason: Now, in the caucus, we revisited all of our previous discussions, and the important thing that came out of that was that, as we had recognized, we did have the possibility that the primary O-ring might be slower to move into the seating position and that was our concern, and that is what we had focused on originally.

The fact that we couldn't show direct correlation with the O-ring temperature was discussed, but we still felt that there was some concern about it being colder.

We then recognized that, if the primary did move more slowly, we could get some blow-by and erosion on the primary. But we had pointed out in that caucus a point that had not come across clearly in our earlier discussions, and that is that we had run tests where we deliberately cut large pieces out of the O-rings to see what the threshold of sealing was, and we found we could go to 125 thousandths of a cut out of the O-ring and it would still seal.

About ten engineers participated in the caucus, along with Mason, Kilminster, C. G. Wiggins (Vice President, Space Division), and Lund. Arnold Thompson and Boisjoly voiced very strong objections to the launch, and the suggestion in their testimony was that Lund was also reluctant to launch:

Mr. Boisjoly: Okay, the caucus started by Mr. Mason stating a management decision was necessary. Those of us who opposed the launch continued to speak out, and I am specifically speaking of Mr. Thompson and myself because in my recollection he and I were the only ones that vigorously continued to oppose the launch. We were attempting to go back and re-review and try to make clear what we were trying to get across, and we couldn't understand why it was going to be reversed. So we spoke out and tried to explain once again the effects of low temperature. Arnie actually got up from his position which was down the table, and walked up the table and put a quarter pad down in front of the table, in front of the management folks, and tried to sketch out once again what his concern was with the joint, and when he realized he wasn't getting through, he just stopped.

I tried one more time with the photos. I grabbed the photos, and I went up and discussed the photos once again and tried to make the point that it was my opinion from actual observations that temperature was indeed a discriminator and we should not ignore the physical evidence that we had observed. Again, I brought up the point that SRM-15 [Flight 51-C, January, 1985] had a 110-degree arc of black grease while SRM-22 [Flight 61-A, October, 1985] had a relatively different amount, which was less and wasn't quite as black. I also stopped when it was apparent that I couldn't get anybody to listen.

Dr. Walker: At this point did anyone else speak up in favor of the launch?

Mr. Boisjoly: No, sir. No one said anything, in my recollection, nobody said a word. It was then being discussed among the management folks. After Arnie and I had our last say, Mr. Mason said we have to make a management decision. He turned to Bob Lund and asked him to take off his engineering hat and put on his management hat. From this point on, management formulated the points to base their decision on. There was never one comment in favor, as I have said, of launching by any engineer or other non-management person in the room before or after the caucus. I was not even asked to participate in giving any input to the final decision charts.

I went back on the net with the final charts or final chart, which was the rationale for launching, and that was presented by Mr. Kilminster. It was hand written on a notepad, and he read from that notepad. I did not agree with some of the statements that were being made to support the decision. I was never asked nor polled, and it was clearly a management decision from that point.

I must emphasize, that I had my say, and I never (would) take (away) any management right to take the input of an engineer and then make a decision based upon that input, and I truly believe that. I have worked at a lot of companies, and that has been done from time to time, and I truly believe that, so there was no point in me doing anything any further than I had already attempted to do. I did not see the final version of the chart until the next day. I just heard it read. I left the room feeling badly defeated, but I felt I did all. I could to stop the launch. I felt that management was under a lot of pressure to launch and they made a very tough decision, but I didn't agree with it.

One of my colleagues who was in the meeting summed it up best. This was a meeting where the determination was to launch, and it was up to us to prove beyond a shadow of a doubt that it was not safe to do so. This is in total reverse to what the position usually is in a preflight conversation or a flight readiness review. It is usually exactly the opposite of that.

Dr. Walker: Do you know the source of the pressure on management that you alluded to?

Mr. Boisjoly: Well, the comments made over the [net] is what I felt, I can't speak for them, but I felt it--I felt the tone of the meeting exactly as I summed up, that we were being put in a position to prove that we should not launch rather than being put in the position and prove that we had enough data to launch. And I felt that very real.

Dr. Walker: These were the comments from the NASA people at Marshall and at Kennedy Space Center?

Mr. Boisjoly: Yes.

Dr. Feynman: I take it you were trying to find proof that the seal would fail?

Mr. Boisjoly: Yes.

Dr. Feynman: And of course, you didn't, you couldn't because five of them didn't, and if you had proved that they would have all failed, you would have found yourself incorrect because five of them didn't fail.

Mr. Boisjoly: That is right. I was very concerned that the cold temperatures would change that timing and put us in another regime, and that was the whole basis of my fighting that night.

After the discussion between Morton Thiokol management and the engineers, a final management review was conducted by Mason, Lund, Kilminster, and Wiggins. Lund and Mason recall this review as an unemotional, rational discussion of the engineering facts as they knew them at that time.

Mr. Lund: We tried to have the telecon, as I remember it was about 6:00 o'clock (MST), but we didn't quite get things in order, and we started transmitting charts down to Marshall around 6:00 or 6:30 [MST], something like that, and we were making charts in real-time and seeing the data, and we were discussing them with the Marshall folks who went along.

We finally got the--all the charts in, and when we got all the charts in I stood at the board and tried to draw the conclusions that we had out of the charts that had been presented, and we came up with a conclusions chart and said that we didn't feel like it was a wise thing to fly.

Question: What were some of the conclusions?

Mr. Lund: I had better look at the chart. Well, we were concerned the temperature was going to be lower than the 50 or the 53 that had flown the previous January, and we had experienced some blow-by, and so we were concerned about that, and although the erosion on the O-rings, and it wasn't critical, that, you know, there had been some little puff go through. It had been caught. There was no real extensive erosion of that O-ring, so it wasn't a major concern, but we said, gee, you know, we just don't know how much further we can go below the 51 or 53 degrees or whatever it was. So we were concerned with the unknown. And we presented that to Marshall, and that rationale was rejected. They said that they didn't accept that rationale, and they would like us to consider some other thoughts that they had.

... Mr. Mulloy said he did not accept that, and Mr. Hardy said he was appalled that we would make such a recommendation. And that made me ponder of what I'd missed, and so we said, what did we miss, and Mr. Mulloy said, well, I would like you to consider these other thoughts that we have had down here. And he presented a very strong and forthright rationale of what they thought was going on in that joint and how they thought that the thing was happening, and they said, we'd like you to consider that when they had some thoughts that we had not considered… So after the discussion with Mr. Mulloy, and he presented that, we said, well, let's ponder that a little bit, so we went off-line to talk about what we—

Question: Who requested to go off-line?

Mr. Lund: I guess it was Joe Kilminster… And so we went off-line on the telecon… So we could have a roundtable discussion here.

Question: Who were the management people that were there?

Mr. Lund: Jerry Mason, Cal. Wiggins, Joe, I, manager of engineering design, the manager of applied mechanics. On the chart.

Before the Commission on February 25, 1986, Mr. Lund testified as follows regarding why he changed his position on launching Challenger during the management caucus when he was asked by Mr. Mason "To take off his engineering hat and put on his management hat":

Chairman Rogers: How do you explain the fact that you seemed to change your mind when you changed your hat?

Mr. Lund: I guess we have got to go back a little further in the conversation than that. We have dealt with Marshall for a long time and have always been in the position of defending our position to make sure that we were ready to fly, and I guess I didn't realize until after that meeting and after several days that we had absolutely changed our position from what we had been before. But that evening I guess I had never had those kinds of things come from the people at Marshall. We had to prove to them that we weren't ready, and so we got ourselves in the thought process that we were trying to find some way to prove to them it wouldn't work, and we were unable to do that. We couldn't prove absolutely that that motor wouldn't work.

Chairman Rogers: In other words, you honestly believed that you had a duty to prove that it would not work?

Mr. Lund: Well, that is kind of the mode we got ourselves into that evening. It seems like we have always been in the opposite mode. I should have detected that, but I did not, but the roles kind of switched …

Supplemental testimony of Mr. Mason obtained in a Commission staff interview is as follows:

Question: Do you recall Mr. Hardy and Mr. Mulloy's comments after--I think after Mr. Kilminster had got done, or Mr. Lund got done presenting the charts? They had some comments. Do you recall—

Mr. Mason: Oh, yes, it was over and over. Hardy said that, "I'm appalled at your recommendation." …

Question: Well, did Mr. Hardy's "appalled" remark and Mr. Mulloy's "can't launch, we won't be able to launch until April" remark, how did that affect your thinking and affect your decision?

Mr. Mason: My personal thinking, I just, you know, it didn't make that much difference .... And the comments that they made, in my view, probably had got more reaction from the engineer[s] at the lower level than they would from the manager[s], because we deal with people, and managers all the time....

Mr. McDonald indicated that during the period of the internal Morton Thiokol caucus he continued to argue for delay with Mulloy, challenging, among other things, the rationale that the rocket motor was qualified down to 40 degrees Fahrenheit. Present were Reinartz, Jack Buchanan, the manager of Morton Thiokol Launch Support Services at Kennedy, and Cecil Houston. McDonald's testimony described that conversation:

Discussion: At this point, what differences of opinion would you expect between managers and engineers about the decision to launch? Which functional area would normally be more in favor of launching? Why?

Mr. McDonald: ... while they were off-line, reevaluating or reassessing this data ... I got into a dialogue with the NASA people about such things as qualification and launch commit criteria. The comment I made was it is my understanding that the motor was supposedly qualified to 40 to 90 degrees. I've only been on the program less than three years, but I don't believe it was. I don't believe that all of those systems, elements, and subsystems were qualified to that temperature.

And Mr. Mulloy said well, 40 degrees is propellant mean bulk temperature, and we're well within that. That is a requirement. We're at 55 degrees for that--and that the other elements can be below that ... that, as long as we don't fall out of the propellant mean bulk temperature. I told him I thought that was asinine because you could expose that large SRM to extremely low temperatures--I don't care if it's 100 below zero for several hours--with that massive amount of propellant, which is a great insulator, and not change that propellant mean bulk temperature but only a few degrees, and I don't think the spec really meant that.

But that was my interpretation because I had been working quite a bit on the filament wound case SRM. It was my impression that the qualification temperature was 40 to 90, and I knew everything wasn't qualified to that temperature, in my opinion. But we were trying to qualify that case itself at 40 to 90 degrees for the filament wound case. I then said I may be naive about what generates launch commit criteria, but it was my impression that launch commit criteria was based upon whatever the lowest temperature, or whatever loads, or whatever environment was imposed on any element or subsystem of the Shuttle. And if you are operating outside of those, no matter which one it was, then you had violated some launch commit criteria.

That was my impression of what that was. And I still didn't understand how NASA could accept a recommendation of fly below 40 degrees. I could see why they took issue with the 53, but I could never see why they would have accepted a recommendation below 40 degrees, even though I didn't agree that the motor was fully qualified to 40. I made the statement that if we're wrong and something goes wrong on this flight, I wouldn't want to have to be the person to stand up in front of board of inquiry and say that I went ahead and told them to go ahead and fly this thing outside what the motor was qualified to. I made that very statement.

Mr. Mulloy's recollections of these discussions are as follows:

Mr. Mulloy: Mr. Kilminster then requested an off-net caucus. It has been suggested, implied, or stated that we directed Thiokol to go reconsider these data. That is not true. Thiokol asked for a caucus so that they could consider the discussions that had ensued and the comments that Mr. Hardy and I and others had made. That caucus, as has been stated, was going to start at that point, and Mr. McDonald interjected into the teleconference. At that point, he made the first comment that he had made during this entire teleconference.

Mr. McDonald testified for quite a while yesterday about his thoughts on this, but he did not say any of them until this point. At that point, he stated that he thought what George Hardy said was a very important consideration, and that consideration was, and he asked Mr. Kilminster to be sure and consider the comment made by George Hardy during the course of the discussions, that the concerns expressed were for primary O-ring blow-by and that the secondary O-ring was in a position to seal during the time of blow-by and would do so before significant joint rotation had occurred. They then went into their caucus, having asked for five minutes—

Mr. Hotz: … It figures quite prominently in the discussion that you were quoted as saying, do you expect us to wait till April to launch?

Mr. Mulloy: Yes, sir.

Dr. Walker: Is that an accurate statement or not?

Mr. Mulloy: It is certainly a statement that is out of context, and the way I read the quote, sir--and I have seen it many times, too many times--the quote I read was: My God, Thiokol, when do you want me to launch, next April? Mr. McDonald testified to another quote that says: You guys are generating new Launch Commit Criteria. Now, both of those I think kind of go together, and that is what I was saying. I don't know whether that occurred during the caucus or subsequent to. I just simply can't remember that.

Mr. Hotz: Well, never mind the timing.

Mr. Mulloy: Well, yes, sir. I'm going to answer your question now. I think those quotes derive from a single thought that may have been expressed by me using some of those words. I have not yet encountered anyone other than those at KSC who heard those words, so I don't believe they were transmitted over the net. The total context I think in which those words may have been used is, there are currently no Launch Commit Criteria [LCC] for joint temperature. What you are proposing to do is to generate a new Launch Commit Criteria on the eve of launch, after we have successfully flown with the existing Launch Commit Criteria 24 previous times. With this LCC, i.e., do not launch with a temperature greater[sic] than 53 degrees, we may not be able to launch until next April. We need to consider this carefully before we jump to any conclusions. It is all in the context, again, with challenging your interpretation of the data, what does it mean and is it logical, is it truly logical that we really have a system that has to be 53 degrees to fly?

At approximately 11 p.m., the Thiokol/NASA teleconference resumed, the Thiokol management stating that they had reassessed the problem, that the temperature effects were a concern, but that the data were admittedly inconclusive. Kilminster read the rationale recommending launch and stated that that was Morton Thiokol's recommendation. Hardy requested that it be sent in writing by telefax both to Kennedy and to Marshall, and it was. The testimony of Mulloy and Hardy regarding the remainder of the teleconference and their rationale for recommending launch follows:

Mr. Mulloy: Okay, sir. At the completion of the caucus, of course, Mr. Kilminster came back on the loop and stated they had assessed all the data and considered the discussions that had ensued for the past couple of hours and the discussions that occurred during their caucus.

Chairman Rogers: Was it a couple of hours?

Mr. Mulloy: Yes, sir. We started at 8:45 and I believe it was probably 11:00 o'clock before he came back on the loop. It was a long discussion. And I must emphasize that I had no knowledge of what interchange occurred during the caucus at Thiokol, because all sites were on mute. We were on mute at KSC. No communications occurred between myself and Mr. Hardy at Huntsville, nor did any communication occur between KSC and Thiokol during that caucus. After Mr. Kilminster made that recommendation, Mr. Reinartz then asked if there were any further comments, and to my recollection there were none. There were no further comments made.

I then asked Mr. Kilminster to send me a copy of his flight readiness rationale and recommendation. The conference was then terminated at approximately 11:15. I have no knowledge of, as has been testified, of Mr. McDonald being asked to sign that documentation. That would have been unusual, because Mr. Kilminster signs all flight readiness documentation.

Now, after the teleconference was complete, Mr. McDonald informed Mr. Reinartz and me that if the Thiokol engineering concern for the effect of cold was not sufficient cause to recommend not launching, there were two other considerations, launch pad ice and recovery area weather. I stated that launch pad ice had been considered by the Mission Management Team—

Chairman Rogers: Excuse me. Could you identify that discussion, where that took place?

Mr. Mulloy: That was after the teleconference was completed, after Mr. Kilminster made his recommendation, after Mr. Reinartz asked "are there any other comments?" There were no other comments on the telecon from anyone .... I stated that launch pad ice had been considered by the Mission Management Team before deciding to proceed and that a further periodic monitoring of that condition was planned. I further stated that I had been made aware of the recovery area weather previously and planned to place a call to Mr. Aldrich and advise him that the weather in the recovery area exceeded the Launch Commit Criteria. As I stated earlier, when you asked what were the Launch Commit Criteria, one of them was that the recovery area weather has limitations on it. The report we had, that Mr. McDonald confirmed, was that we were outside of those limits. Now, I must point out that that is not a hard Launch Commit Criteria. That is an advisory call, and the LCC so states that. It does require that we discuss the condition.

So at about 11:30 p.m., Mr. Cecil Houston established a teleconference with Mr. Aldrich and Mr. Sestile at KSC. I informed Mr. Aldrich that the weather in the recovery area could preclude immediate recovery of the SRBs, since the ships were in a survival mode and they were moving back toward Cape Kennedy at about three knots, and the estimate provided to us by Mr. Sestile was that they would be probably 40 miles from the SRB impact area at the time of launch, at 9:38; and then, continuing at three knots, it was going to be some period of time before they could get back and locate the boosters.

The concern I had for that was not loss of the total booster, but loss of the main parachutes for the booster, which are separated at water impact, and loss of the frustum of the boosters, which has the drogue parachute on it, which comes down separately, because with the 50 knot winds we had out there and with the kind of sea states we had, by the time the recovery ships got back out there, there was little probability of being able to recover those. I informed Mr. Aldrich of that, and he decided to proceed with the launch after that information. I did not discuss with Mr. Aldrich the conversations that we had just completed with Morton Thiokol.

Chairman Rogers: Could you explain why?

Mr. Mulloy: Yes, sir. At the time, and I still consider today, that was a Level III issue, Level III being an SRB element or an external tank element or Space Shuttle main engine element or an Orbiter. There was no violation of Launch Commit Criteria. There was no waiver required in my judgment at that time and still today. And we work many problems at the Orbiter and the SRB and the External Tank level that never get communicated to Mr. Aldrich or Mr. Moore. It was clearly a Level III issue that had been resolved.

… There were 27 full-scale seal tests with an O-ring groove damage tolerances, damage in the grooves and damage tolerance on O-rings. And then there were two cold gas tests. And these data were presented on the night of the 27th. All of that was at ambient temperature. And then we did discuss what is a development qualification motor experience range, and that is shown on the chart. We had experience everywhere from 40 to 85 degrees. There then were data presented on two cold gas tests at 30 degrees, where the O-ring was pressurized at the motor pressurization rate at 30 degrees, which would indicate that an O-ring would operate before joint rotation at 30 degrees.

Dr. Ride: Was that actually in a joint?

Mr. Mulloy: No, it is not. It is a full-scale O-ring, full-scale groove, in a scaled test device, where the pressurize rate on that O-ring is zero to 900 psi[pounds per square inch] in 600 milliseconds at a temperature of 30 degrees.

Dr. Walker: You would say, then, the O-ring was qualified to a temperature of 30 degrees? Would that be an accurate statement?

Mr. Mulloy: The day that we were looking at it, on the 27th, these two tests that we did indicated that it would perform at 30 degrees under the motor pressurization rate before the joint rotated.

Dr. Walker: What about, let's consider the putty and the O-ring, because that is really the system that responds to the pressure surge. What temperature was the putty/O-ring system qualified to?

Mr. Mulloy: The lowest that I'm aware of--and we're still flushing this out, because this is kind of what we talked about on the 27th, but the lowest that I'm aware of is the 40-degree test on one of the development motors.

Dr. Walker: And, of course, during those tests the putty was modified before the test. The putty was not just laid up and then the seal made. The putty was then smoothed out or some attempt was made to remove the volcanoes, I think.

Mr. Mulloy: Because the horizontal assembly caused that.

Now, there's one other significant point on this chart that we did discuss, that we didn't have the quantities on the 27th, and I mentioned this earlier. We have 150 case segment proof tests, with a large number of joints with a simulation of a cold O-ring. That is the 90 durometer with a.275, and that was at about 35 degrees.

So those are the certification data that we kind of discussed, all of which we didn't discuss. The two cold gas tests we did, the segment proof test we did, the development and qualification motor test we did, as a basis for understanding what we could expect to happen at colder temperatures on the joints.

Mr. Hardy testified as follows:

Mr. Hardy: At the teleconference on the evening of January 27, 1986, Thiokol engineering personnel in Utah reviewed charts that had been datafaxed to Huntsville and KSC participants just prior to the beginning of the conference. Now, I am not going to repeat a lot of what you have already heard, but I will give you some of my views on the whole matter.

The presentations were professional in nature. There were numerous questions and answers. There was a discussion on various data and points raised by individuals at Thiokol or at Marshall or at Kennedy. I think it was a rather full discussion. There were some 14 charts presented, and as has been mentioned earlier, we spent about two, two and a half hours reviewing this. To my knowledge, anyone who desired to make a point, as a question or express a view was in no way restrained from doing so.

As others have mentioned, I have heard this particular teleconference characterized as a heated discussion. I acknowledge that there were penetrating questions that were asked, I think, from both, from all people involved. There were various points of view and an interpretation of the data that was exchanged. The discussion was not, in my view, uncharacteristic of discussions on many flight readiness issues on many previous occasions. Thiokol engineering concluded their presentation with recommendation that the launch time be determined consistent with flight experience to date, and that is the launch with the O-ring temperatures at or greater than 53 degrees Fahrenheit.

Mr. Kilminster at Thiokol stated … to the best of my recollection, that with that engineering assessment, he recommended we not launch on Tuesday morning as scheduled. After some short discussion, Mr. Mulloy at KSC summarized his assessment of the data and his rationale with that data, and I think he has testified to that.

Mr. Reinartz, who was at KSC, asked me for comment, and I stated I was somewhat appalled, and that was referring specifically to some of the data or the interpretation of some of the data that Thiokol had presented with respect to its influence on the joint seal performance relative to the issue under discussion, which specifically was the possibility that the primary seal may take longer to actuate and therefore to blow by the primary seal. The blow-by of the primary seal may be longer, and I am going to elaborate on that a little further in this statement.

Then I went on to say that I supported the assessment of data presented essentially as summarized by Mr. Mulloy, but I would not recommend launch over Thiokol's objections. Somewhere about this time, Mr. Kilminster at Utah stated that he wanted to go off the loop to caucus for about five minutes. I believe at this point Mr. McDonald, the senior Thiokol representative at KSC for this launch suggested to Mr. Kilminster that he consider a point that I think I had made earlier, that the secondary O-ring is in the proper position to seal if blow-by of the primary O-ring occurred.

I clearly interpreted this as a somewhat positive statement of supporting rationale for launch … The status of the caucus by Thiokol lasted some 30, 35 minutes. At Huntsville during this Thiokol caucus, we continued to discuss the data presented. We were off the loop, we were on mute. We were around a table in small groups. It was not an organized type discussion. But I did take that opportunity to discuss my assessment and understanding of the data with several of my key advisors, and none of us had any disagreement or differences in our interpretation of what we believe the data was telling us with regard to the primary issue at hand.

When Thiokol came back on line, Mr. Kilminster reviewed rationale that supported proceeding with the launch and so recommended Mr. Reinartz asked if anyone in the loop had a different position or disagreed or something to that effect, with the Thiokol recommendation as presented by Mr. Kilminster. There were no dissenting responses. The telecon was terminated shortly after, and I have no knowledge of any subsequent events or discussions between personnel at KSC or at Thiokol on this matter.

At about 5:00 a.m. on January 28, a discussion took place among Messrs. Mulloy, Lucas, and Reinartz in which Mulloy reported to Lucas only that there had been a discussion with Thiokol over their concerns about temperature effects on the O-rings, and that it had been resolved in favor of launch. The following testimony of Mr. Mulloy and Dr. Lucas recount that discussion:

General Kutyna: … Larry, let me follow through on that, I am kind of aware of the launch decision process, and you said you made the decision at your level on this thing. If this were an airplane, an airliner, and I just had a two-hour argument with Boeing on whether the wing was going to fall off or not, I think I would tell the pilot, at least mention it. Why didn't we escalate a decision of this importance?

Mr. Mulloy: I did, sir.

General Kutyna: You did?

Mr. Mulloy: Yes, sir.

General Kutyna: Tell me what levels above you.

Mr. Mulloy: As I stated earlier, Mr. Reinartz, who is my manager, was at the meeting, and on the morning, about 5:00 o'clock in the operations support room where we all were I informed Dr. Lucas of the content of the discussion.

General Kutyna: But this is not in the launch decision chain.

Mr. Mulloy: No, sir. Mr. Reinartz is in the launch decision chain, though.

General Kutyna: And is he the highest level in that chain?

Mr. Mulloy: No. Normally it would go from me to Mr. Reinartz to Mr. Aldrich to Mr. Moore.

Discussion: Obviously, a few people are attempting to get the right information to the right people, but there are some trying to obstruct the flow of information. Construct a taxonomy of the launch proponents and opponents. Continue to add to the list as additional names surface during the testimony. Include the organization to which each participant belong.

Dr. Lucas' testimony is as follows:

Chairman Rogers: Would you please tell the Commission when you first heard about the problem of the O-rings and the seals insofar as it involves launch 51-L? And I don't want you to go way back, but go back to when you first heard. I guess it was on January 27th, was it?

Dr. Lucas: Yes, sir. It was on the early evening of the 27th, I think about 7:00 p.m., when I was in my motel room along with Mr. Kingsbury. And about that time, Mr. Reinartz and Mr. Mulloy came to my room and told me that they had heard that some members of Thiokol had raised a concern about the performance of the Solid Rocket Boosters in the low temperature that was anticipated for the next day, specifically on the seals, and that they were going out to the Kennedy Space Center to engage in a telecon with the appropriate engineers back at Marshall Space Flight Center in Huntsville and with corresponding people back at the Wasatch division of Thiokol in Utah. And we discussed it a few moments and I said, fine, keep me informed, let me know what happens.

Chairman Rogers: And when was the next time you heard something about that?

Dr. Lucas: The next time was about 5:00 a.m. on the following morning, when I went to the Kennedy Space Center and went to the launch control center. I immediately saw Mr. Reinartz and Mr. Mulloy and asked them how the matter of the previous evening was dispositioned.

Chairman Rogers: You had heard nothing at all in between?

Dr. Lucas: No, sir.

Chairman Rogers: So from 8:00 o'clock that evening until 5:00 o'clock in the morning, you had not heard a thing?

Dr. Lucas: It was about 7:00, I believe, sir. But for that period of time, I heard nothing in the interim…

Chairman Rogers: … And you heard Mr. Reinartz say he didn't think he had to notify you, or did he notify you?

Dr. Lucas: He told me, as I testified, when I went into the control room, that an issue had been resolved, that there were some people at Thiokol who had a concern about the weather, that that had been discussed very thoroughly by the Thiokol people and by the Marshall Space Flight Center people, and it had been concluded agreeably that there was no problem, that he had a recommendation by Thiokol to launch and our most knowledgeable people and engineering talent agreed with that. So from my perspective, I didn't have--I didn't see that as an issue.

Chairman Rogers: And if you had known that Thiokol engineers almost to a man opposed the flight, would that have changed your view?

Dr. Lucas: I'm certain that it would.

Chairman Rogers: So your testimony is the same as Mr. Hardy's. Had he known, he would not have recommended the flight be launched on that day.

Dr. Lucas: I didn't make a recommendation one way or the other. But had I known that, I would have then interposed an objection, yes.

Chairman Rogers: I gather you didn't tell Mr. Aldrich or Mr. Moore what Mr. Reinartz had told you?

Dr. Lucas: No, sir. That is not the reporting channel. Mr. Reinartz reports directly to Mr. Aldrich. In a sense, Mr. Reinartz informs me as the institutional manager of the progress that he is making in implementing his program, but that I have never on any occasion reported to Mr. Aldrich.

Chairman Rogers: And you had subsequent conversations with Mr. Moore and Mr. Aldrich prior to the flight and you never mentioned what Mr. Reinartz had told you?

Dr. Lucas: I did not mention what Mr. Reinartz told me, because Mr. Reinartz had indicated to me there was not an issue, that we had a unanimous position between Thiokol and the Marshall Space Flight Center, and there was no issue in his judgment, nor in mine as he explained it to me.

Chairman Rogers: But had you known, your attitude would have been totally different?

Dr. Lucas: Had I had the advantages at that time of the testimony that I have heard here this week, I would have had a different attitude, certainly.

Chairman Rogers: In view of the fact that you were running tests to improve the joint, didn't the fact that the weather was so bad and Reinartz had told you about the questions that had been raised by Thiokol, at least, didn't that cause you serious concern?

Dr. Lucas: I would have been concerned if Thiokol had come in and said, we don't think you should launch because we've got bad weather.

Chairman Rogers: Well, that's what they did, of course, first. That is exactly what they did. You didn't know that?

Dr. Lucas: I knew only that Thiokol had raised a concern.

Chairman Rogers: 'Did you know they came and recommended against the launch,' is the question.

Dr. Lucas: I knew that I was told on the morning of the launch that the initial position of some members of Thiokol--and I don't know who it was--had recommended that one not launch with the temperature less than 53 degrees Fahrenheit.

Chairman Rogers: And that didn't cause you enough concern so you passed that information on to either Mr. Moore or Mr. Aldrich?

Dr. Lucas: No, sir, because I was shown a document signed by Mr. Kilminster that indicated that that would not be significant, that the temperature would not be--that it would be that much lower, as I recall it.

It is clear that crucial information about the O-ring damage in prior flights and about the Thiokol engineers' argument with the NASA telecon participants never reached Jesse Moore or Arnold Aldrich, the Level I and II program officials, or J. A. (Gene) Thomas, the Launch Director for 51-L. The testimony of Aldrich describes this failure of the communication system very aptly:

Dr. Feynman: ... have you collected your thoughts yet on what you think is the cause--I wouldn't call it of the accident but the lack of communication which we have seen and which everybody is worried about from one level to another? ...

Mr. Aldrich: Well, there were two specific breakdowns at least, in my impression, about that situation. One is the situation that occurred the night before the launch and leading up to the launch where there was a significant review that has been characterized in a number of ways before the Commission and the Commission's Subpanels and the fact that that was not passed forward.

And I can only conclude what has been reported, and that is that the people responsible for that work in the Solid Rocket Booster project at Marshall believed that the concern was not of a significance that would be required to be brought forward because clearly the program requirements specify that critical problems should be brought forward to Level II and not only to Level II but through myself to Level I.

The second breakdown in communications, however, and one that I personally am concerned about is the situation of the variety of reviews that were conducted last summer between the NASA Headquarters Organization and the Marshall Organization on the same technical area and the fact that that was not brought through my office in either direction--that is, it was not worked through--by the NASA Headquarters Organization nor when the Marshall Organization brought these concerns to be reported were we involved.

And I believe that is a critical breakdown in process and I think it is also against the documented reporting channels that the program is supposed to operate to.

Now, it in fact did occur in that matter. In fact, there is a third area of concern to me in the way the program has operated. There is yet one other way that could have come to me, given a different program structure. I'm sure you've had it reported to you as it has been reported to me that in August or I think or at least at some time late in the summer or early fall the Marshall SRB project went forward to procure some additional SRM casings to be machined and new configurations for testing of the joints.

Now it turns out that the budget for that kind of work does not come through my Level II office. It is worked directly between the Marshall Center in NASA headquarters and there again had I been responsible for the budget for that sort of work, it would have to come through me, and it would have been clear that something was going on here that I ought to know about. And so there are three areas of breakdown, and I haven't exactly answered your question. But I have explained it in the way that I best know it and--well, I can say a fourth thing.

There was some discussion earlier about the amount of material that was or was not reported on O-ring erosion in the FRRs [Flight Readiness Review] and I researched the FRR back reports and also the flight anomaly reports that were forwarded to my center--to my office--by the SRB[Solid Rocket Booster] project and as was indicated, there is a treatment of the SRM O-ring erosion, I believe, for the STS 41-C FRR, which quantifies it and indicates some limited amount of concern.

The next time that is mentioned, I believe it is the STS 51-E, FRR in January 1985 or early in February, and that indicates, again, a reference to it but refers back to the 41-C as the only technical data. And then from there forward the comment on O-ring erosion only is that there was another instance and it is not of concern. Clearly the amount of reporting in the FRR is of concern to me, but in parallel with that, each of the flight anomalies in the STS program are required to be logged and reviewed by each of the projects and then submitted through the Level II system for formal closeout.

And in looking back and reviewing the anomaly close-outs that were submitted to Level II from the SRB project, you find that O-ring erosion was not considered to be an anomaly and, therefore, it was not logged and, therefore, there are not anomaly reports that progress from one flight to the other. Yes, that is another way that that information could have flagged the system, and the system is set up to use that technique for flagging. But if the erosion is classified as not an anomaly, it then is in some other category and the system did not force it in that direction. None of those are very focused answers, but they were all factors.

Chairman Rogers asked four key officials about their knowledge of the Thiokol objections to launch:

Chairman Rogers: ... By way of a question, could I ask, did any of your gentlemen prior to launch know about the objections of Thiokol to the launch?

Mr. Smith [Kennedy Space Center Director]: I did not.

Mr. Thomas [Launch Director]: No, sir.

Mr. Aldrich [Shuttle Program Director]: I did not.

Mr. Moore [Associate Administrator for Space Flight]: I did not.

Additionally, in further testimony J. A. (Gene) Thomas commented on the launch.

Mr. Hotz: … Mr. Thomas, you are familiar with the testimony that this Commission has taken in the last several days on the relationship of temperature to the seals in the Solid Rocket Booster?

Mr. Thomas: Yes, sir, I have been here all week.

Mr. Hotz: Is this the type of information that you feel that you should have as launch Director to make a launch decision?

Mr. Thomas: If you refer to the fact that the temperature according to the Launch Commit Criteria should have been 53 degrees, as has been testified, rather than 31, yes, I expect that to be in the LCC. That is a controlling document that we use in most cases to make a decision for launch.

Mr. Hotz: But you are not really very happy about not having had this information before the launch?

Mr. Thomas: No, sir. I can assure you that if we had had that information, we wouldn't have launched if it hadn't been 53 degrees.

PRESIDENTIAL COMMISSION FINDINGS

The following are the findings or conclusions of the Presidential Commission based on the testimony heard.

1. There was a serious flaw in the decision making process leading up to the launch of flight 51-L. A well-structured and managed system emphasizing safety would have flagged the rising doubts about the Solid Rocket Booster joint seal. Had these matters been clearly stated and emphasized in the flight readiness process in terms reflecting the views of most of the Thiokol engineers and at least some of the Marshall engineers, it seems likely that the launch of 51-L might not have occurred when it did.
2. The waiving of launch constraints appears to have been at the expense of flight safety. There was no system which made it imperative that launch constraints and waivers of launch constraints be considered by all levels of management.
3. There was a propensity of management at Marshall to contain potentially serious problems and to attempt to resolve them internally rather than communicate them forward. This tendency is altogether at odds with the need for Marshall to function as part of a system working toward successful flight missions, interfacing and communicating with the other parts of the system that work to the same end.
4. The Presidential Commission concluded that the Thiokol Management reversed its position and recommended the launch of 51-L, at the urging of Marshall and contrary to the views of its engineers is order to accommodate a major customer.

Discussion: Evaluate your own operating environment and see if you can find situations that closely identify with the four problem areas listed above. How could each problem be resolved? Should a communication matrix have been developed, monitored, and enforced by NASA for the shuttle mission?

MANAGEMENT DECISION AMBIGUITIES

Another source of problem for the Challenger was the unusually cold weather on the night before the launch. Reaction control system heaters on the Orbiter were activated and the Solid Rocket Booster recovery batteries were checked and found to be functioning within specifications. There were no serious concerns regarding the External Tank. The freeze protection plan for the launch pad was implemented, but the results were not what had been anticipated. The freeze protection plan usually involves completely draining the water system. However, this was not possible because of the imminent launch of 51-L. In order to prevent pipes from freezing, a decision was made to allow water to run slowly from the system. This had never been done before, and the combination of freezing temperatures and stiff winds caused large amounts of ice to form below the 240-foot level of the fixed service structure including the access to the crew emergency egress slide wire baskets. Ice also was forming in the water trays beneath the vehicle.

These conditions were first identified by the ice team at approximately 2:00 a.m. on January 28 and were assessed by management and engineering throughout the night, culminating with a Mission Management Team meeting at 9:00 a.m. At this meeting, representatives for the Orbiter prime contractor, Rockwell International, expressed their concern about what effects the ice might have on the Orbiter during launch. Rockwell had been alerted about the icing conditions during the early morning and was working on the problem at its Downey, California, facility.

During Commission hearings, the president of Rockwell's Space Transportation Systems Division, Dr. Rocco Petrone, and two of his vice presidents, Robert Glaysher and Martin Cioffoletti, all described the work done regarding the ice conditions and the Rockwell position at the 9:00 a.m. meeting with regard to launch. Dr. Petrone had arrived at Kennedy on Friday, January 24. On Monday the 27th he left to return to Rockwell's facility in California, but Glaysher and Cioffoletti remained at Kennedy. Dr. Petrone testified that he first heard about the ice at 4:00 a.m. Pacific Standard Time. He explained what followed:

> I had gotten up and went to the support room to support this launch. We have people monitoring consoles, and I checked in, and they told me there was a concern, and when I arrived at about 4:30, I was informed we were working the problem with our aerodynamicist and debris people, but very importantly, we would have to make an input to Kennedy for a meeting scheduled at 6:00 o'clock our time and 9:00 o'clock Florida time.
>
> We had approximately an hour of work to bring together. The work had been underway when I arrived and was continuing.
>
> At that time I got on the phone with my Orbiter program managers just to discuss background of where we were, how things stood, and what their concerns were locally. They described what they knew in Florida, and we also in Downey did television input, and we could see some of the ice scenes that were shown here this morning.
>
> We arrived through a series of meetings to a top level discussion at approximately 5:30 Pacific Standard Time, from which we drew the following

> conclusions: Ice on the mobile launcher itself, it could be debris. We were very concerned with debris of any kind at the time of launch. With this particular ice, one, could it hit the Orbiter? There was wind blowing from the west. That appeared not to be so, that it wouldn't hit the Orbiter but would land on the mobile launcher. The second concern was what happens to that ice at the time you light your liquid-fuel engines, the SSMEs, and would it throw it around and ricochet and potentially hit the Orbiter.
>
> The third aspect is the one that has been discussed here of aspiration, what would happen when the large solid rocket motors (SRMs) motors ignite and in effect suck in air, referred to as aspiration, and ice additionally would come down, how much unknown.
>
> The prime thing we were concerned about was the unknown base line. We had not launched in conditions of that nature, and we just felt we had an unknown.
>
> I then called my program managers over in Florida at 5:45 (PST) and said we could not recommend launching from here, from what we see. We think the tiles would be endangered, and we had a very short conversation. We had a meeting to go through, and I said let's make sure that NASA understands that Rockwell feels it is not safe to launch, and that was the end of my conversation.

Mr. Glaysher, who was at Kennedy, came to the center at approximately 7:45 a.m. EST. He conferred with Rockwell's Chief Engineer as well as the Vice President of Engineering, Dr. John Peller, at Rockwell's Downey Plant. At 9:00 a.m., after the ice debris team had reported back from the pad inspection, Glaysher was asked for Rockwell's position on launch. He discussed aspiration effects, the possible ricochet of ice from the fixed service structure, and what the ice resting on the mobile launch platform would do at ignition. Glaysher said he told the Mission Management Team when it met at 9:00 a.m. that the ice was an unknown condition, and Rockwell was unable to predict where the ice would go or the degree of potential damage to the Orbiter thermal protection system if it were struck by the ice. He testified that his recommendation to NASA was:

> My exact quote--and it comes in two parts. The first one was, Rockwell could not 100 percent assure that it is safe to fly which I quickly changed to Rockwell cannot assure that it is safety to fly…

Rockwell's other vice president at Kennedy, Martin Cioffoletti, described the concern about ice in a slightly different manner:

Mr. Cioffoletti: Similarly, I was called in and told about the problem and came into the 6:00 o'clock meeting which you heard about a few minutes ago, and at the conclusion of that meeting I spoke with Mr. Dick Kohrs, the deputy program manager from Johnson Space Flight Center, and he asked if we could get the Downey folks to look at the falling ice and how I might reverse toward the vehicle, and also, did we have any information on aspiration effects.

So I did call back to Downey and got the John Peller folks working on that problem, and they did, as you saw from Charlie Stevenson's sketches, predict that the ice would travel only about halfway to the vehicle, freefalling ice carried by the winds. So we felt that ice was not a problem. However, it would land on the mobile launch platform. That we considered a problem. We also investigated the aspiration data base we had, and we had seen the aspiration effect on previous launches where things were pulled into the SRB[Solid Rocket Booster] hole after ignition, but we had never seen anything out as far as the fixed surface tower. So we felt in fact it was an unknown. We did not have the data base to operate from an aspiration effect.

At the 9:00 o'clock meeting, I was asked by Arnie Aldrich, the program manager, to give him the results of our analysis, and I essentially told him what I just told you and felt that we did not have a sufficient data base to absolutely assure that nothing would strike the vehicle, and so we could not lend our 100 percent credence, if you will, to the fact that it was safe to fly … I said I could not predict the trajectory that the ice on the mobile launch platform would take at SRB ignition.

Chairman Rogers: But I think NASA's position probably would be that they thought that you were satisfied with the launch. Did you convey to them in a way that they were able to understand that you were not approving the launch from your standpoint?

Mr. Cioffoletti: I felt that by telling them we did not have a sufficient data base and could not analyze the trajectory of the ice, I felt he understood that Rockwell was not giving a positive indication that we were for the launch.

After Cioffoletti's testimony at the Commission hearings, Dr. Petrone was pressed for a more detailed description of Rockwell's launch recommendation:

General Kutyna: Dr. Petrone, you've got a lot more experience than I have in this business, but the few launch conferences that I have been on the question is very simple. Are you go or are you no-go for launch, and "maybe" isn't an answer. I hear all kinds of qualifications and cautions and considerations here.

Did someone ask you are you go or no-go? Was that not asked?

Dr. Petrone: At this particular meeting, as far as--and I was not in Florida, and so I cannot answer that. It had been done at earlier meetings. This was a technical evaluation of a series of problems, and we talked about debris hitting the TPS[thermal protection system] and the tiles, and the long series of reviews that we had done that morning and all led us to a conclusion that they were not safe to fly. And we transmitted that to program managers along with the technical evaluation quickly of why we had arrived at that. So much of it is how the question gets raised because earlier we had aspiration work, ricochet work, a number of things which we did, and then we came up with our recommendation.

Chairman Rogers: And your recommendation now you say it was, it was unsafe to fly?

Dr. Petrone: Correct, sir.

Two things are apparent from the Rockwell testimony. First, Rockwell did not feel it had sufficient time to research and resolve the ice on the pad problem. Second, even though there was considerable discussion about ice, Rockwell's position on launch described above was not clearly communicated to NASA officials in the launch decision chain during the hours preceding 51-L's launch.

At a meeting with Commission investigators on March 4, 1986, at Kennedy, HoraceLamberth, NASA director of Shuttle Engineering, said he did not interpret Rockwell's position at the 9:00 a.m. Mission Management Team meeting on January 28 as being "no-go." Lamberth said the language used by Rockwell was "we can't give you 100 percent assurance" but there was no feeling in his mind that Rockwell was voicing a no-go recommendation. "It just didn't come across as the normal Rockwell no-go safety of flight issues come across." This conclusion is confirmed in part by an interview of Dr. John Peller, Rockwell's Vice President of Engineering, who was assigned the ice problem early Tuesday morning. Dr. Peller, in describing a telephone conversation with the Johnson Director of Engineering, Tom Moser, stated:

Dr. Peller: That was a call from Tom Moser to me, in which he asked again to understand my concerns. And I just repeated the same concerns. And he asked, "Did I think that it was likely that the vehicle would take safety critical damage?"

And I said, "From the possibility that the vehicle would take safety critical damage," I said, "there's a probability in a sense that it was probably an unlikely event, but I could not prove that it wouldn't happen…"

…I never used the words "no-go" for launch. I did use the words that we cannot prove it is safe. And normally that's what we were asked to do. We were unable to do that in this particular case, although it was a strange case, that we normally don't get involved in.

Arnold Aldrich, NASA Mission Management Team Leader, described NASA's view of the ice situation and his recollection of Rockwell's position. He said that on Tuesday morning the mission management team did a detailed analysis of the ice on the fixed service structure. Representatives from the ice team, Rockwell, and the directors of Engineering (Horace Lamberth) and the Orbiter project (Richard Colonna) all considered the problem. Aldrich reported the discussion as follows:

> Following the discussion of the acceptability of the ice threat to the Orbiter, based upon the conditions described in detail of the fixed service structure--and more of that you've seen here portrayed well this morning--I asked the NASA managers involved for their position on what they felt about the threat of that to the Orbiter.
>
> Mr. Lamberth reported that KSC (Kennedy Space Center) engineering had calculated the trajectories, as you've heard, of the falling ice from the fixed

> service structure east side, with current 10-knot winds at 300 degrees, and predicted that none of this ice would contact the Orbiter during its ignition or launch sequence; and that their calculations even showed that if the winds would increase to 15 knots, we still would not have contact with the Orbiter.
>
> Mr. Colonna, Orbiter project manager, reported that similar calculations had been performed in Houston by the mission evaluation team there. They concurred in this assessment. And further, Mr. Colonna stated that, even if these calculations were significantly in error, that it was their belief that falling ice from the fixed service structure, if it were in fact to make its way to the Orbiter, it would only be the most lightweight ice that was in that falling stream, and it would impact the Orbiter at a very oblique angle.
>
> Impacts of this type would have very low probability of causing any serious damage to the Orbiter, and at most would result in post-flight turnaround repairs.
>
> At this point I placed a phone call to Mr. Moser that I had previously mentioned, director of Engineering at the Johnson Space Center, who was in the mission evaluation room, and he confirmed the detailed agreement with Mr. Lamberth's and Mr. Colonna's position….
>
> And both Mr. Lamberth and Mr. Colonna reported that their assessment was that the time it took for the ice to fall, to hit the Orbiter and to rebound, and the location of the fixed service structure on the MLP[mobile launch platform] would not cause that ice in their view to be a concern to rebound and come up and impact the rear end of the Orbiter.
>
> Following these discussions, I asked for a position regarding proceeding with the launch. Mr. Colonna, Mr. Lamberth, and Mr. Moser all recommended that we proceed.
>
> At that time, I also polled Mr. Robert Glaysher, the vice president, Orbiter project manager, Rockwell International STS Division, and Mr. Marty Cioffoletti, Shuttle Integration Project Manager, Rockwell International STS Division. Mr. Glaysher stated--and he had been listening to this entire discussion and had not been directly involved with it, but had been party to this the whole time.
>
> His statement to me as best I can reconstruct it to report to you at this time was that, while he did not disagree with the analysis that JSC[Johnson Space Center] and KSC has reported, that they would not give an unqualified go for launch as ice on the launch complex was a condition which had not previously been experienced, and thus this posed a small additional, but unquantifiable, risk. Mr. Glaysher did not ask or insist that we not launch, however.
>
> At the conclusion of the above review, I felt reasonably confident that the launch should proceed.

In addition to Rockwell's input, Mr. Aldrich also had reports from other contractors and the ice, frost and debris team at the 9:00 session. Ice on the vehicle assembly appeared to be of no concern; sheet ice in the noise suppression trays had been broken up and removed; as previously noted the ice team reported that there was ice on the fixed service structure between 95 feet above ground and 215 feet; no ice above

255 feet. The north and west sides had large amounts of ice and icicles. The final assessment was made that the ice on the fixed service structure would not strike or damage the Orbiter tiles or the vehicle assembly during ignition or ascent, owing to the considerable horizontal distance between the service structure and the vehicle assembly. The decision was made to launch pending a final ice team review of the launch complex in order to assess any changes in the situation. This inspection was completed following the Mission Management Team meeting and the ice team report indicated no significant change.

ADDITIONAL COMMISSION FINDINGS

The testimonies on the ice problem were reviewed by the commission and their findings are as presented below:

1. An analysis of all of the testimony and interviews establishes that Rockwell's recommendation on launch was ambiguous. The commission finds it difficult, as did Mr. Aldrich, to conclude that there was a no-launch recommendation. Moreover, all parties were asked specifically to contact Aldrich or Moore about launch objectives due to weather. Rockwell made no phone calls or further objections to Aldrich or other NASA officials after the 9:00 Mission Management Team meeting and subsequent to the resumption of the countdown.
2. The Commission is also concerned about the NASA response to the Rockwell position at the 9:00 a.m. meeting. While it is understood that decisions have to be made in launching a Shuttle, the Commission is not convinced Levels I and II appropriately considered Rockwell's concern about the ice. However, ambiguous Rockwell's position was, it is clear that they did tell NASA that the ice was an unknown condition. Given the extent of the ice on the pad, the admitted unknown effect of the SRM and Space Shuttle Main Engines ignition on the ice, as well as the fact that debris striking the Orbiter was a potential flight safety hazard, the Commission finds the decision to launch questionable under those circumstances. In this situation, NASA appeared to be requiring a contractor to prove that it was not safe to launch, rather than proving it was safe. Nevertheless, the Commission has determined that the ice was not a cause of the 51-L accident and does not conclude that NASA's decision to launch specifically overrode a no-launch recommendation by an element contractor.
3. The Commission concluded that the freeze protection plan for launch pad 39B was inadequate. The Commission believes that the severe cold and presence of so much ice on the fixed service structure made it inadvisable to launch on the morning of January 28, and that margins of safety were whittled down too far.

Additionally, access to the crew emergency slide wire baskets was hazardous due to ice conditions. Had the crew been required to evacuate the orbiter on the launch pad, they would have been running on an icy surface. The Commission believes the crew should have been made aware of the situation, and based on the seriousness of the condition, greater consideration should have been given to delaying the launch.

Discussion:

1. As mentioned above, it seems launch officials wanted a contractor to furnish a proof for "no-go" rather than a proof for "go." Does this qualify as "management by crisis?"
2. The Triple C Principle (see Chapter 5) recommends an integrated approach to project communication, cooperation, and coordination. How could the Triple C be instituted for a complex project in order to avoid the problems presented in this case study?

POST INVESTIGATION DEVELOPMENTS

Some of the developments following the Challenger accident and investigations include:

1. The NASA official, who gave final approval to launch Challenger, Jesse W. Moore, resigned from NASA.
2. It was announced in June, 1986 that a redesign of the booster rocket joints would incorporate full redundancy so that a single failure would not jeopardize future space flights. The new joint would be rated as a "criticality-1R" item. The "R" in the rating implies that there is a redundant capability.
3. The Reagan administration, in August 1986, ordered NASA to get out of the commercial launching business. This marked the termination of a two-decade struggle by NASA to run a space shuttle service that would pay for itself. The order was based on the belief that space services could be provided better and cheaper by the private sector.
4. Critics called for a new supplier of shuttle booster rockets. It was revealed in August 1986 that Morton Thiokol Corporation skipped three of seven mandatory government safety inspections of O-ring seals in Challenger's right-hand solid booster rocket.
5. In October, 1986, an independent panel of scientific and engineering experts endorsed NASA's decision to continue testing its redesigned shuttle booster rockets in the same horizontal position as before the Challenger accident. The panel was organized by the National Research Council at the recommendation of the Presidential Commission. The commission had urged NASA to consider vertical tests in order to duplicate actual launch configuration as closely as possible.
6. Robert Crippen, the astronaut assigned the job of revamping shuttle management, recommended that NASA should learn to analyze the full history and trend of spaceship problems. He said the short-term troubleshooting by NASA "failed to put together the whole story" of the booster rocket.
7. Several of the families of the astronauts who died in the Challenger accident filed lawsuits against NASA and the shuttle contractors. In particular,

the family of Michael J. Smith (pilot) filed a $15 million negligence claim on July 2, 1986. The families of Judith A. Resnik (mission specialist) and Ronald E. McNair (mission specialist) also filed independent lawsuits.

8. In December, 1986, The Justice Department announced that it has reached confidential out-of-court settlements with the families of four of the seven astronauts that died in the accident. The families were those of Christa McAuliffe (school teacher), Francis R. Scobee (mission commander), Ellison S. Onizuka (mission specialist), and Gregory B. Jarvis (payload specialist).
9. As of January, 1987, about $1.2 million had been donated to a fund established for the education, health, and support of the 11 children of the Challenger crew.
10. It was revealed for the first time in August, 1986 that internal NASA memorandums had warned that the O-ring seals could leak and cause a catastrophe. This information had not been provided to the Presidential Commission that investigated the Challenger accident.
11. In October, 1986, a congressional committee says NASA may not have the technical and scientific expertise to conduct the Space Shuttle Program. The House Science and Technology Committee promised to conduct an in-depth review of NASA's technical ability during the 1987 fiscal year.
12. NASA officials announced in January, 1987 that the agency is creating a nap plan for launch managers so that they could get adequate rest during the hectic pace of shuttle flight preparations.
13. In May, 1987, retired astronaut Walter Schirra recommended an agenda for the space program. He called for a space program manned by high-tech astronauts dedicated to no-frills missions. He urged NASA to delay and totally redesign the next space Shuttle.
14. On January 10, 1987, NASA announced the selection of the next Challenger crew consisting of previous space fliers. It was the first time that a space shuttle crew would be comprised entirely of astronauts who had flown in space before.

REFERENCE

Badiru, Adedeji B. (1996), *Project Management in Manufacturing and High Technology Operations*, 2nd edition, John Wiley & Sons, New York (Chapter 11).

Appendix A: 2024 Progress Report on UN's SDGs

SUMMARY

In accordance with General Assembly resolution 70/1, the present report provides a global overview of progress made since 2015 toward the Sustainable Development Goals using inputs from more than 50 international and regional organizations. The data contained herein are based on the global indicator framework developed by the Inter-Agency and Expert Group on Sustainable Development Goal Indicators, adopted by the General Assembly on July 6, 2017 (see Assembly resolution 71/313).

INTRODUCTION

1. As we begin the second half of our journey to 2030, signs of a determined, sustained global comeback have yet to emerge. This year's report reveals that only 17 percent of the SDGs targets are on track to be achieved, nearly half are showing minimal or moderate progress, and progress on over a third has stalled or even regressed.
2. The early years of SDG implementation saw slow but steady progress on some critical SDG indicators, but since 2019, national efforts have faced severe global headwinds. The COVID-19 pandemic, a growing number of conflicts, geopolitical and trade tensions, and the ever-worsening effects of climate change have combined to leave the SDGs in peril. Massive shortcomings in our global economic and financial systems, coupled with historical injustices and a weakening of global solidarity, mean that developing countries and the world's most vulnerable people are bearing the brunt.
3. The pandemic and other factors led to 23 million more people living in extreme poverty and 123 million more suffering from hunger in 2022 compared to 2019, widening the gap in per capita income growth between the poorest and richest countries. While some health targets have improved, overall progress in global health has slowed since 2015. Progress on education—the foundation upon which so much rests—remains of grave concern, with only 58% of students worldwide achieving minimum proficiency in reading by the end of primary school and one in five young people neither in education, training nor employment. The 2030 Agenda's commitment to achieving gender equality remains a distant objective, with limited progress on achieving gender parity in public life and managerial roles and persistently high levels of violence against women and girls.
4. In the area of climate and biodiversity, despite some reductions in greenhouse gas emissions in developed countries, greenhouse concentrations hit

record highs in 2022, with real-time data in 2023 indicating a continued rise. Carbon dioxide levels have surged to 150% above pre-industrial levels. Public funding for oil, coal, and gas production and consumption more than doubled from 2021 to 2022 and tripled since 2015, hindering progress toward a net-zero transition. Ocean acidification is on the rise and will persist if CO_2 emissions continue to climb. Species extinction risk is worsening, with a 12% decline in the aggregate Red List Index since 1993.

5. All the while, violent organized crime and illicit arms trafficking continue to plague numerous countries, posing paramount security threats. And with the Russian invasion of Ukraine now in its third year, a brutal war taking place in Gaza, and the conflict in Sudan yet to show signs of easing, the number of forcibly displaced people worldwide has reached an unprecedented 110 million. Between 2022 and 2023, civilian casualties in armed conflict increased by 72%.
6. While Official Development Assistance (ODA) reached another new high of $223.7 billion in 2023 (corresponding to 0.37% of DAC members' combined GNI) and FDI flows reached $1.37 trillion, the overall sustainable development grants to developing countries fell in 2022 and debt levels—while falling slightly—remain, exceptionally high and the annual investment gap for developing countries to achieve the SDGs stands at $4 trillion.
7. Despite these challenges, there remain glimmers of hope. This report demonstrates that progress is being made on issues such as reducing the global infant mortality rate, the incidence of HIV infections and the cost of remittances, and increasing access to water and sanitation, energy, and mobile broadband. In addition, behind the global trends, many countries are pursuing SDG transformation with real determination and they are making tangible progress.
8. What's more, governments remain united behind the 2030 Agenda. This was one of the key takeaways from the SDG summit in September 2023. It is essential that we protect this commitment by taking the critical steps needed to deliver transformative progress between now and 2030, at much greater speed and scale.
9. Firstly, if we have to achieve the SDGs, we need peace. The conflicts and violence that are causing untold death, destruction, displacement, and disruption across the world must end and a new period of international relations that respects the principles of the UN Charter must begin.
10. Secondly, if our commitment to leaving no one behind is to mean anything, we must unlock much greater financing and fiscal space for developing countries and secure a more equitable, representative, and effective international financial system. Over the course of the next 18 months, through the Summit of the Future, the work of the G20 and international financial institutions, COP29, and the Fourth International Conference on Financing for Development, we must deliver concrete changes to the international policies, instruments, and institutions that greatly impact the flow of development finance.
11. Thirdly, we must double down on those areas that can unlock transformative progress across the goals. Key transitions around energy, food, and

digital connectivity, underpinned by expanded access to social protection, decent jobs, education, and skills for the future, are essential for rewiring economies to combat the triple planetary crisis and to reduce inequalities within and between countries. Achieving these objectives requires not only strengthened institutions and policy frameworks but also robust investment pathways, and strengthened support from the UN development system, multilateral and regional development banks, as well as private investors. Plans for these transitions can also inform updated national climate plans by 2025 that cover the whole economy and align with the 1.5 degree limit. Implementation efforts must be done hand in hand with a continued push to tackle the root causes of gender inequality.

12. As this report lays bare, international cooperation and decisive action are urgently needed to recapture the spirit and resolve that gave birth to the 2030 Agenda nine years ago. Our shared aspirations for a more peaceful, prosperous, and sustainable future for all people and the planet itself hangs precariously in the balance. Only through renewed multilateral efforts and the unwavering commitment of all stakeholders can we uplift the millions currently being left behind and accelerate toward the achievements of the SDGs by 2030.

SDG Progress in 2024

13. The 2024 progress assessment reveals the world is severely off track to achieve the 2030 Agenda. As illustrated in Figures A.1 and A.2, out of 135 targets with trend data and additional insights from custodian agencies, only 17% are progressing as expected to be achieved by 2030. Nearly half (48%) exhibit moderate to severe deviations from the desired trajectory, with 30% showing marginal progress and 18% indicating moderate progress. Alarmingly, 18% have stagnated, and 17% have regressed below the 2015 baseline levels.
14. The above assessment is affected by ongoing challenges pertaining to SDG data availability and timeliness. Overall, as illustrated in Figure A.3, commendable progress has been made in improving data availability for SDG monitoring when comparing the 2019 global SDG database and the 2024 database. Back in 2016, when the indicator framework was initially agreed, only around one-third of the indicators had good data coverage (i.e., data

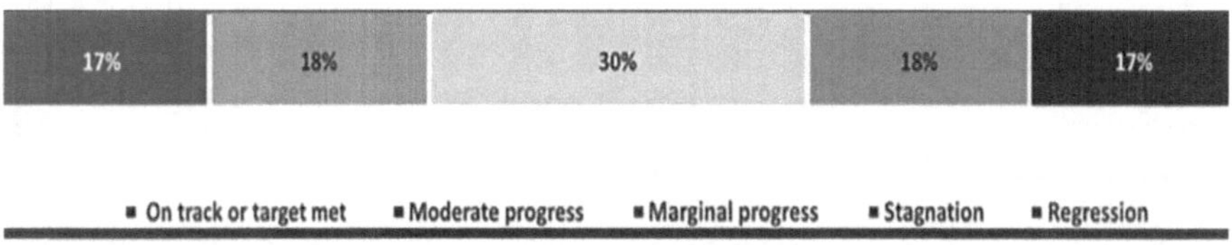

FIGURE A.1 Overall progress assessment across targets with trend data, 2024, or the latest data.

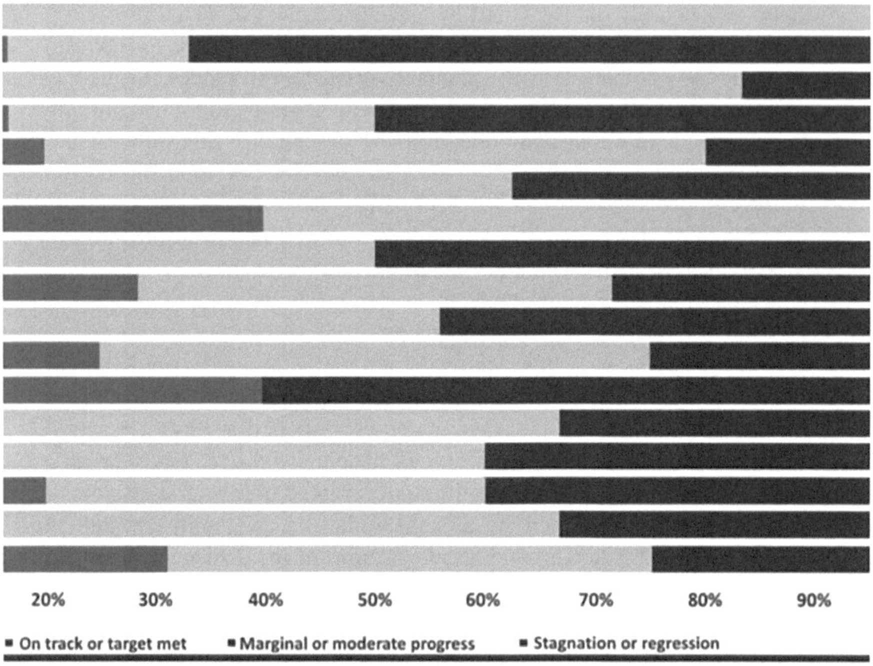

FIGURE A.2 Progress assessment for the 17 Goals based on assessed targets with trend data, by Goal, 2024 or the latest data.

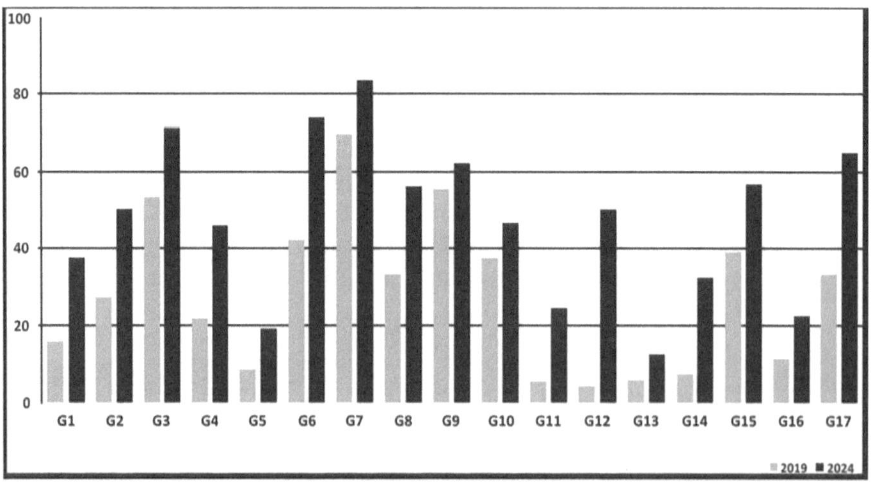

FIGURE A.3 Proportion of countries or areas with available data (at least two data points since 2015), comparing the 2019 database and 2024 database, by Goal (percentage).

was available for more than 50% countries) and a further 39% of the SDG indicators lacked internationally established methodologies or standards. Today 68% of indicators have good data coverage and all 231 indicators have a well-established and internationally agreed methodology since 2020. In addition, good trend data is available for 51% of indicators (at least two data points since 2015) in more than half of all countries.

15. The variance in country data coverage plays out very differently across the goals, with significant data gaps in priority development areas such as gender equality (Goal 5), climate action (Goal 13), and peace, justice, and strong institutions (Goal 16). Moreover, data timeliness remains a challenge, with approximately one-third of the indicators lacking data for the last three years. The lack of timely data hampers policymakers' ability to make informed decisions and course corrections.
16. Recognizing the relevance of data to guide progress across all SDGs, a high-impact initiative on the power of data was launched at the SDG summit in September 2023. This initiative aims to deliver the data needed to achieve the SDGs by mobilizing political support and investments in national data systems, rallying around one national data strategy, and strengthening partnerships, collaboration, and coordination. The expanding initiative currently comprises 15 countries across Africa, Asia, Latin America, and the Caribbean. More countries and partners are encouraged to join this important quest to ensure every government has the data it needs to both track and drive SDG progress.
17. The remainder of this section provides an overview of progress under each goal with brief assessments of selected targets. Global and regional data, along with progress assessments for all targets and indicators for which information is available can be found in the statistical annex (Figure A.3).

Goal 1. End Poverty in All Its Forms Everywhere

18. Global efforts to eradicate extreme poverty have faced significant setbacks due to the COVID-19 pandemic and a series of major shocks during 2020–22. The pandemic caused extreme poverty to increase in 2020 for the first time in decades, reversing global progress by three years. Recovery has been uneven, with low-income countries lagging behind. With the ongoing polycrisis, ending poverty by 2030 appears increasingly out of reach, particularly in regions that lack the fiscal capacity to cope with economic stresses.
    - **Target 1.1**: Extreme poverty levels returned to pre-pandemic levels in most countries by 2022, except in low-income countries where recovery has been slower. In 2022, 9% of the world's population or 712 million people were living in extreme poverty, an increase of 23 million people compared to 2019. If current trends continue, 590 million people, or 6.9% of the world's population will still live in extreme poverty by 2030.

      The share of the world's working population living in poverty has steadily decreased, from 8.4% in 2015 to 6.9% in 2023. However, nearly 241 million workers globally were still living in extreme poverty in 2023 and little positive change is expected in 2024.

- **Target 1.2**: Though data covering the period of COVID-19 are limited, the pandemic is likely to have slowed progress made in halving national poverty rates. Given historical trends, less than 30% of countries worldwide will have halved poverty by 2030.
- **Target 1.3**: In 2023, only 28.2% of children globally received child cash benefits, compared to 22.1% in 2015, leaving 1.4 billion children aged 0–15 without coverage. To guarantee at least a basic level of social protection for all children, upper- and lower-middle-income countries would need to invest an additional US$98.1 billion and US$88.8 billion, respectively, while low-income countries would require an additional US$59.6 billion.
- **Target 1.5**: Economic losses due to disasters remained stubbornly high in recent years and showed no sign of alleviation. Between 2015 and 2022 direct economic loss has been reported to exceed more than $115 billion per year worldwide, which amounted to 0.3% of the GDP of the reporting countries.
- **Target 1.a**: Data from approximately 100 countries show that the proportion of total government spending on essential services (education, health, and social protection) is approximately 50%, with an average of 60% among advanced economies and 40% among emerging market and developing economies. While this indicator trends slightly upward for both groups over the past two decades, the gap between them remains stable at approximately 20 percentage points.

Goal 2. End Hunger, Achieve Food Security and Improved Nutrition, and Promote Sustainable Agriculture

19. Globally, hunger persists with nearly 1 in 10 of the world's population facing it in 2022, while 2.4 billion people experienced moderate to severe food insecurity. In the same year, nearly 60 percent of countries worldwide saw significant increases in food prices due to conflicts and disrupted supply chains. Achieving zero hunger requires intensified efforts to transform food systems toward sustainability, resilience, and equity. Furthermore, accelerating improvements in diets, nutrition, health, and hygiene is crucial to meeting the SDG target of halving the number of children suffering from chronic undernutrition.
    - **Target 2.1**: After a sharp increase following the COVID-19 pandemic, global hunger stabilized at around 9.2% of the population from 2021 to 2022. Between 691 and 783 million people faced hunger in 2022. Considering the midrange (735 million), 122 million more people faced hunger in 2022 than in 2019, when the prevalence stood at 7.9%. Additionally, an estimated 29.6% of the global population—2.4 billion people—were moderately or severely food insecure in 2022.
    - **Target 2.2**: Globally in 2022, an estimated 22.3% of children under age five (148 million) were affected by stunting, down from 24.6% in 2015 and 26.3% in 2012 (baseline year of WHO nutrition targets). Based on current trends, one out of five (19.5%) children under age five will be

affected by stunting in 2030. Overweight affected 37.0 million children under age five (or 5.6%) and wasting affected 45 million (or 6.8%) in 2022.

- **Target 2.3**: The income gap between small-scale and non-small-scale food producers remains significant. In 95% of countries with available data, the average annual income of small-scale producers is less than half of non- small-scale producers. Among small-scale food producers, units headed by men typically generate higher incomes compared to those headed by women.
- **Target 2.a**: In 2022, global public expenditures reached $36 trillion, of which $749 billion went toward agriculture—an all-time high. Agriculture represented 2.1% of total government expenditure, a recovery from the pandemic, and is only marginally below the 2019 level. Government expenditure on agriculture relative to the agriculture sector's contribution to GDP, measured in terms of the agriculture orientation index, declined from the 2015 baseline of 0.50 to 0.43 in 2021, but rebounded to 0.48 in 2022.
- **Target 2.b**: Total notified annual agriculture export subsidy outlays fell from its peak of $ 6.7 billion in 1999 to $33 million in 2022. As of January 1, 2024, only least developed countries (LDCs) and net food importing developing countries are allowed to use certain forms of export subsidies.
- **Target 2.c**: In 2022, the share of countries facing moderately to abnormally high food prices reached a new record high of 58.1% after falling by more than half in 2021 from the previous historical peak of 48% in 2020. The 2022 share represented a nearly four-fold increase from 2015 to 2019 average levels of 15.2%. Major disruption to logistics and food supply chains, following the breakout of the war in Ukraine, resulted in higher food and energy prices, particularly during the first half of 2022.

Goal 3. Ensure Healthy Lives and Promote Well-Being for All at All Ages

20. Achieving global health goals faces significant challenges, with progress slowing since 2015 in areas like maternal mortality, premature deaths from major non-communicable diseases, and access to essential healthcare. Inequalities persist, especially among vulnerable populations, exacerbated by the climate crisis. To meet the SDG 3 targets by 2030, substantial investment and focus are needed to address these challenges, including tackling inequality and environmental factors. Urgent action is required to protect vulnerable groups and regions with high disease burdens.
    - **Target 3.1**: The global maternal mortality ratio marginally declined from 227 maternal deaths per 100,000 live births in 2015 to 223 in 2020. Reaching the global maternal mortality rate target of 70 per 10,000 live births will require an annual rate of reduction of 11.6% between 2021 and 2030. Two regions, Sub- Saharan Africa and Southern Asia, accounted for around 87% (249,000) of the estimated global maternal deaths in 2020. Globally, skilled birth attendance rose from 80% in

2015 to 86% in 2023. There are, however, significant regional disparities, particularly in Sub-Saharan Africa where the rate was just 73% in 2023.

- **Target 3.2**: In 2022, global under-five deaths was 4.9 million, down from 9.9 million in 2000 and 6.0 million in 2015. The under-five mortality rate fell to 37 deaths per 1,000 live births in 2022—51% lower than in 2000 and a 14% decline since 2015. Similarly, the global neonatal mortality rate fell to 17 deaths per 1,000 live births in 2022, a 44% and 12% decrease from the 2000 and 2015 levels, respectively.
- **Target 3.3**: Mixed progress is observed toward the SDG target of ending communicable diseases
  a. There were an estimated 1.3 million new HIV infections in 2022, 27% fewer than in 2015, and 38% fewer than in 2010. Increased access to HIV treatment has averted almost 20.8 million AIDS-related deaths in the past three decades.
  b. The reported global number of people newly diagnosed with TB was 7.5 million in 2022, the highest since 1995. On the other hand, the annual number of people who died from TB decreased in 2022 after two consecutive years of increase due to the COVID-19 pandemic.
  c. In 2022, there were an estimated 249 million malaria cases globally, exceeding the pre-pandemic level of 233 million in 2019.
  d. In 2022, 1.62 billion people required interventions and care for neglected tropical diseases (NTDs), a 26.1% of decline from 2010. As of December 2023, 50 countries, territories and areas have eliminated at least one NTD.
- **Target 3.5**: Global drug-related treatment coverage has decreased from approximately 11% in 2015 to under 9% in 2022. Alarmingly, treatment coverage for women consistently lags behind that for men across all regions. In 2022, over 13% of men with drug use disorders received treatment globally, while less than 6% of women did. Moreover, data on treatment coverage for alcohol use disorders vary widely, ranging from a mere 0.3% to a maximum of 14% in reporting countries.
- **Target 3.7**: The proportion of women of reproductive age (aged 15–49 years) who have their need for family planning satisfied with modern methods increased slightly from 76.5% to 77.6% between 2015 and 2024. This corresponds to an increase of 75 million women of reproductive age using modern methods since 2015. The adolescent birth rate has globally declined from 47.2 births per 1,000 girls and women aged 15–19 years in 2015 to 40.7 in 2024.
- **Target 3.8**: The proportion of the population not covered by essential health services decreased by about 15% between 2000 and 2021, with minimal progress made after 2015. In 2021, about four and a half billion people were not covered by essential health services.
- **Target 3.a**: In 2022, the global prevalence of current tobacco use among the population aged 15+ was estimated at 20.9%. This translates to around 1.25 billion adult tobacco users in the world. The prevalence

has declined since 2015 when it was 23.9%, and the number of users has decreased by 50 million.

- **Target 3.b**: Coverage of the third dose of vaccine protecting against diphtheria, tetanus, and pertussis (DTP-3) recovered to 84% in 2022, an improvement from 81% in 2021 but still below the 2019 level of 86%. In 2022, 20.5 million children remained vulnerable to vaccine-preventable diseases. The current 2-dose measles vaccine coverage of 74% is insufficient to prevent outbreaks. HPV vaccine coverage declined significantly during the pandemic, but 2022 saw the first encouraging signs of recovery in vaccination with 15% full scheduled coverage among girls.
- **Target 3.c**: While a recent study shows that the projected global shortage of health workers by 2030 has reduced from 18 million to 10 million, the aging of the population induces an increased health need and further widens this gap. An additional 1.8 million health workers are needed in 54 countries (mostly from high-income countries) just to maintain the current age-standardized density of health workers.

Goal 4. Ensure Inclusive and Equitable Quality Education and Promote Lifelong Learning Opportunities for All

21. SDG4 is a key enabler of most other SDGs. Unfortunately, global progress in education has not been fast enough. Only 58% of students worldwide achieved at least the minimum proficiency level in reading at the end of primary schooling in 2019. A large share of countries is moving backward in learning outcomes at the end of lower secondary school. Improvement in upper secondary completion rate has slowed since 2015. Some regions, including Sub-Saharan Africa, are facing teacher shortages, high student–teacher ratios, and inadequate training and lack of professional development opportunities for teachers. Accelerating progress toward SDG 4 should be prioritized as it will have a catalytic impact on achieving the overall 2030 Agenda.
    - **Target 4.1**: Completion rates in primary and lower secondary level education continue on an upward curve while the percentage of young people completing upper secondary school increased from 53% in 2015 to 59% in 2023. This increase, however is at a slower pace relative to progress in the preceding eight- year period and such improvements do not always result in positive learning outcomes. Between 2018 and 2022, based on learning outcomes at the end of lower secondary school across 81 OECD and partner countries, mean performance in mathematics fell by a record 15 points while in reading fell 10 points. However, reading and mathematics scores had been declining for these countries prior to 2015, suggesting that COVID-19 explains only part of the decline. A complex set of factors affect the education systems of upper-middle-income and high-income countries.
    - **Target 4.2**: Data from 76 mainly low- and middle-income countries from 2015 to 2023 shows that approximately two-thirds of young children are developmentally on track, with no significant gender

differences. However, there are wide variations among countries and regions. In 2022, globally, seven out of ten children participated in organized learning one year before reaching official primary school age. Despite a 1.5 percentage point decline during the COVID-19 pandemic, participation levels have started to recover and return to pre-pandemic levels.

- **Target 4.3**: In countries with recent data, around one-sixth of individuals aged 15–64 have participated in education and training. Participation is significantly higher among youth aged 15–24 compared to those aged 24–55, with an average participation rate of nearly 50% across most regions. However, less than 3% of older adults aged 25–55 engage in education and training in most regions.
- **Target 4.5**: Socioeconomic disparities are prevalent in education, affecting various indicators. Global and regional parity ratios may hide gender inequalities within countries, disadvantaging either girls or boys. Disparities based on location or household wealth are more pronounced, with rural or less affluent families facing greater challenges. These gaps widen at higher education levels, leading to increased dropout rates and fewer opportunities for those from disadvantaged backgrounds.
- **Target 4.a**: Only half of all primary schools have the basic infrastructures and materials to provide an adequate schooling experience to pupils with disabilities and one in five primary schools globally does not have single-sex sanitation facilities. On average, 44% of primary schools, 56% of lower secondary schools and 69% of upper secondary schools had access to internet in 2022, almost double the rates in 2021. At the upper secondary level, 91% of schools have access to electricity, 81% have computers for pedagogical use and 69% of schools are connected to the internet.
- **Target 4.c**: In 2022, 15% of teachers were still not trained according to national minimum standards in their country. No progress on this indicator has been observed since 2015. Regions differ substantially: in Sub-Saharan Africa, only 70% of teachers meet their national minimum qualification requirement to be considered trained.

Goal 5. Achieve Gender Equality and Empower All Women and Girls

22. Progress toward gender equality is clearly off track. Harmful practices like child marriage and female genital mutilation are decreasing, but not fast enough to keep pace with population growth. Gender parity, especially in women's participation in public life and managerial roles, remains distant. At the current rate, achieving gender parity in managerial positions will take 176 years. Many women still lack control over their sexual and reproductive health, and violence against women persists. Urgent action is needed to challenge biased social norms, eliminate harmful practices, and change discriminatory laws. Increasing women's leadership roles and investments in gender equality are crucial at national, regional, and global levels.

- **Target 5.1**: Based on data collected in 2022 in 120 countries, 55% of the countries lacked non-discrimination laws that prohibit direct and indirect discrimination against women and half of the countries continued to lack quotas for women in the national parliament.
- **Target 5.3**:
  - Globally, around 640 million girls and women were married before age 18, with India accounting for one-third. While progress has reduced child marriage, still, one in five young women (19%) are married before 18, down 22% in 2013. Despite averting 68 million child marriages in this period, the pace isn't sufficient to eliminate the practice by 2030.
  - Over 230 million girls and women worldwide are estimated have undergone female genital mutilation as of 2024, an increase of 30 million compared to the last estimate in 2016.
- **Target 5.4**: On an average day, women spend about 2.5 times as many hours in unpaid domestic and care work as men, according to the latest data between 2000 and 2022.
- **Target 5.5**: Progress in women's political participation remains slow. As of January 1, 2024, women held 26.9% of seats in national parliaments worldwide, marking a modest increase of 4.6 percentage points since 2015 (22.3%). In local governments, the participation was higher at 35.5% as of January 1, 2023. While global employment sees nearly 40% representation from women, they only held 27.5% of management positions in 2022, a notable decrease from 28.5% in 2021.
- **Target 5.a**: In one in three of the 49 countries with data, less than 50% of women and men have ownership or secure rights over agricultural land. Furthermore, there is a notable disparity between women and men's agricultural land ownership, with men owning land at least twice as often as women in almost half of the countries. Legal protections for women's land rights are low or not existent in approximately 58% of the reporting countries.
- **Target 5.b**: In 2023, 78% of the global population aged 10 and above owned a mobile phone. Women were approximately 8% less likely to own a mobile phone than men in 2023, a decrease from 10% in 2020.
- **Target 5.c**: Data from 105 countries and areas for the period 2018–2021 show that 26% of countries globally have comprehensive systems to track and make public allocations for gender equality, 59% have some features of a system, and 15% do not have minimum elements of these systems.

Goal 6. Ensure Availability and Sustainable Management of Water and Sanitation for All

23. None of the SDG 6 targets are on track to be met. As of 2022, 2.2 billion people were without access to safely managed drinking water and 3.5 billion lacked access to safely managed sanitation. Between 2002 and 2021 droughts affected more than 1.4 billion people. As of 2022, roughly half of

the world's population experienced severe water scarcity for at least part of the year, while one-quarter faced 'extremely high' levels of water stress. Climate change worsens these issues, posing significant risks to social stability. While transboundary river and lake basins are home to approximately 40% of the world's population and are shared by 153 countries, less than a fifth of these countries have operational arrangements for cooperation in place for all their transboundary waters. Coordinated global action on integrated water management, addressing the impacts of climate change on water resources, and better management of shared waters are key to restoring focus on the Goal.

- **Targets 6.1 and 6.2**: Between 2015 and 2022, the population using safely managed drinking water increased from 69% to 73%, the population using safely managed sanitation increased from 49% to 57% and the population using basic hygiene services increased from 67% to 75%. Achieving universal coverage of safely managed drinking water, safely managed sanitation, and basic hygiene services by 2030 will require respective increases of 6x, 5x, and 3x in the current global rates of progress.
- **Target 6.3**: Among the 73 countries reporting on both total wastewater generation and total wastewater treatment in 2022, 76% of total wastewater flows received at least some treatment. Of the 42 countries that specified the level of treatment, 60% of total wastewater flows were safely treated (i.e., at least secondary treatment). In all world regions, many rivers, lakes and aquifers are still in good condition—as of 2023, 56% of water bodies assessed in 120 countries have good water quality. However, countries that implement the most extensive monitoring programs show that water quality is degrading since 2017.
- **Target 6.4**: Water use efficiency rose from \$17.4/m^3 in 2015 to \$20.8/m^3 in 2021, a 19% increase. At the global level, the water stress level reached an average level of 18.6% in 2021 but it hides large regional variations as some regions in the world show critical water stress levels which may compromise their economic and social development. Globally, water stress level has slightly increased by 3% from 2015 to 2021.
- **Target 6.5**: Global progress on implementing integrated water resources management remains slow—49% in 2017, 54% in 2020, 57% in 2023, not on track to reach the 2030 target (91%). Only 43 out of 153 countries have operational arrangements in place for 90% or more of their shared transboundary waters (rivers, lakes, and aquifers), and more than 20 countries have no operational arrangements in place for any of their transboundary waters. Since 2020 only around ten new arrangements for transboundary water cooperation have been adopted.
- **Target 6.6**: The world has experienced a net increase in permanent surface water between 2005 and 2022, much of which resulted from climate change and reservoir filling. While this global net gain is positive, it is important to recognize that the data also include losses of permanent water in some regions linked to droughts and increased water demands.

- **Target 6.a**: ODA disbursements to the water sector increased between 2021 and 2022 by 10% to $8.5 billion, reversing a declining trend seen over the past five years.

Goal 7. Ensure Access to Affordable, Reliable, Sustainable, and Modern Energy for All

24. In 2022, global electricity access declined for the first time in a decade, primarily due to disruptions from COVID-19 and the Ukraine conflict. Despite improvements in energy intensity and renewable energy growth, international financial flows for clean energy in developing countries remain insufficient. At the current rate, 660 million people will still lack electricity and 1.8 billion will not have access to clean cooking by 2030. To achieve universal access to energy by 2030, we need to expedite electrification efforts, boost investments in renewable energy, enhance energy efficiency, and establish supportive policies and regulatory frameworks.
    - **Target 7.1**:
        - In 2022, global electricity access remained at 91%, but the number without access increased by 10 million from 2021 to 685 million people. Factors such as COVID-19 and the Ukraine conflict disrupted progress. Projections suggest that by 2030, 660 million will still lack electricity. Sub-Saharan Africa accounted for 83% of deficit in 2022. However, Central and Southern Asia reduced their gap from 235 million in 2015 to 33 million in 2022. Annual progress slowed to 0.4% between 2020 and 2022, requiring a 1.08% increase until 2030 to meet the target.
        - In 2022, around 74% of the world used clean fuels for cooking. Yet, about 2.1 billion still relied on polluting fuels and technologies such as charcoal, coal, crop waste, dung, kerosene, and wood. The global access deficit decreased from 36% to 26% since 2015. However, current trends suggest a 21% shortfall in achieving universal access by 2030, leaving 1.8 billion without access to clean cooking by 2030.
    - **Target 7.2**: In 2021 the global share of renewable sources in total final energy consumption stood at 18.7%. Excluding traditional use of biomass, the share of modern renewable sources rose gradually from 10% in 2015 to 12.5% in 2021. The electricity sector led the charge with renewables, contributing 28.2% to total final electricity consumption. However, insufficient progress in the heat and transport sectors underscores the need for stronger conservation measures and policy actions. Tripling world's installed renewable energy generation agreed at the COP28 is an important step aligning with the SDG7.
    - **Target 7.3**: In 2021, the primary energy intensity improved by 0.8%, falling below both the 1.2% five-year average and the SDG 7.3 target of 2.6%. To meet the 2030 target, annual improvements must now average around 4%. The robust economic recovery in 2021 led to the largest annual rise in energy consumption in 50 years, exceeding 5%.

This surge was driven by a shift toward energy- intensive industries and the resurgence of other demand sectors after lockdowns were lifted.

- **Target 7.a**: In 2022, international public financial flows supporting clean energy in developing countries rose to $15.4 billion, a 25% increase from 2021 but still half of the 2016 peak of $28.5 billion. However, in 2023, it was anticipated a decrease in global five-year average flows by $450 million. The decreasing trend in these flows may hinder SDG 7 achievement, especially for LDCs, Land Locked Developing Countries (LLDCs), and Sudden Infant Death Syndrome (SIDS).
- **Target 7.b**: Installed renewable energy capacity is on the rise worldwide, reaching 424 watts per person globally in 2022. Developed nations averaged 1073 watts per person, while developing countries averaged 293 watts per person. This represents an 8.5% increase from 2021, maintaining a steady compound annual growth rate of 8.1% over five-year periods.

Goal 8. Promote Sustained, Inclusive and Sustainable Economic Growth, Full and Productive Employment, and Decent Work for All

25. Progress toward SDG 8 faces challenges from the COVID-19 aftermath, trade tensions, rising debts in developing nations, conflicts, and geopolitical strains, collectively threatening global economic growth. While labor markets have shown resilience, uneven pandemic recovery, the declining protection of labor rights and emerging vulnerabilities erode social justice prospects. The report foresees a worsening labor market outlook, with higher unemployment and sluggish growth in 2024, exacerbating income inequality and jeopardizing equitable pay for women and decent work for young people. Achieving SDG 8 mandates policies fostering economic growth with a focus on social justice and inclusive employment.
    - **Target 8.1**: After a sharp 3.9% decline in 2020 due to the COVID-19 pandemic, the global economy rebounded with a 5.3% increase in real GDP per capita in 2021. However, growth slowed to 2.2% in 2022 and is forecasted to slow down further to 1.0% in 2023, before picking up slightly to 1.8% and 1.5% in 2024 and 2025, respectively. For LDCs, real GDP growth dropped from 5.1% in 2019 to 0.7% in 2020, then recovered to 3.8% in 2021 and 4.6% in 2022. Growth is expected to rise to 4.4% and 5.5% in 2023 and 2024. However, it is projected to slow down to 4.9% in 2025.
    - **Target 8.2**: Productivity growth stagnated in 2022 and 2023, remaining below 0.5%. This sluggish trend stands in stark contrast to the pre-pandemic period of 2015 to 2019, where the rate exceeded 1.5%. The pandemic sharply interrupted this trend, with 2020 registering a marked decline as output fell faster than employment—though this was fully offset by a short-lived rebound of productivity during 2021. The recent slow productivity growth poses a risk to economic development and living standards, given its crucial role as a driver of growth.

- **Target 8.3**: In 2023, over 2 billion workers globally were employed informally, accounting for 58.0% of the global workforce. This figure is expected to see a marginal decrease to 57.8% in 2024. The decline of the informality rate by less than a percentage point since 2015 is far too slow for widespread formalization to occur anytime soon.
- **Target 8.5**: In 2023, the global unemployment rate not only rebounded, dipping below its pre-pandemic level, but also achieved a historic low since 2000, settling at 5.1%. However, projections indicate a slight increase in global unemployment in 2024, with approximately 2 million more individuals unemployed, leading to a 5.2% unemployment rate. Women and youth continue to experience higher unemployment rates compared to their male and adult counterparts worldwide and across most regions.
- **Target 8.6**: In 2023, the global NEET (not in education, employment, or training) rate for young people was 21.7%, showing a significant decrease since 2020 and nearing the 2015 baseline of 21.8%. This rate is expected to persist through 2025. There is a critical need to intensify initiatives aimed at reducing NEET rates among youth, especially focusing on young women. Globally, young women are still more than twice as likely as young men to be NEET.
- **Target 8.8**:
  - Globally, many workers face significant risks in their workplaces, and work accidents remain prevalent. In 11 out of 93 countries with data, more than ten work-related fatalities per 100,000 workers were reported. Additionally, in half of the 96 countries with available data, the number of non-fatal injuries per 100,000 workers exceeded 641.
  - From 2015 to 2022, the global average level of national compliance with labor rights declined by 7%. This decline is observed across both developed and developing countries and has become more pronounced in all regions since 2020. Recent data indicates that ongoing crises have led to an increase in violations of labor rights in practice and, alarmingly, by violations of fundamental civil liberties of workers, employers, and their organizations.
- **Target 8.9**: In 2022, tourism rebounded to 82% of its 2019 level, contributing 3.1% to global GDP. Lifted travel restrictions and pent-up demand drove this recovery, but regional differences persist. Oceania excluding Australia and New Zealand and SIDS still faced challenges, with tourism economic performance at 68% and 43% of pre-pandemic levels, respectively.
- **Target 8.10**: Since 2015, global access to finance has increased, but recent years show changes in access methods. The number of ATMs per 100,000 adults declined from 64.6 in 2015 to 63.9 in 2022, while commercial bank branches dropped from 15.0 to 13.7 per 100,000 adults. The COVID-19 pandemic has accelerated the shift toward digital

financial access. This reflects a global decline in bank branches, except in Central Asia, Southern Asia, and Sub-Saharan Africa.

- **Target 8.10**: The Global Findex 2021 reports that despite 74% of adults having a financial account, 41% lack financial resilience. Financial resilience means being unable to access extra funds equivalent to 5% of their country's gross national income within 30 days of a financial shock like a health emergency or job loss. This varies by region, with South Asia being the least financially resilient (only 32%) and East Asia the most (77%).
- **Target 8.b**: In 2023, less than half of the reporting countries (36 out of 87) had implemented a national strategy for youth employment. About one-third of these countries possess a strategy but lack clear evidence of its implementation, while one-fifth are in the process of developing one.

Goal 9. Build Resilient Infrastructure, Promote Inclusive and Sustainable Industrialization, and Foster Innovation

26. Since 2022, the manufacturing sector has faced stagnation, attributed to geopolitical instability, inflation, logistical challenges, rising energy costs, and a broader global economic slowdown. Globally, manufacturing's share in employment has regressed. While there has been progress in reducing CO_2 intensity in manufacturing, it falls short of 2030 target values. To expedite progress toward SDG 9, efforts should prioritize accelerating the green transition, strategically prioritizing sectors, and addressing inequalities in digital and innovation sectors.
    - **Target 9.2**:
        - The manufacturing sector rebounded strongly in 2021 post-COVID, but growth has plateaued at around 2.7% since 2022, expected to continue in 2024. Despite this, global manufacturing value added per capita rose by 16% from 2015 to 2023, reaching $1,922 per capita. Regional gaps are stark, with Europe and Northern America hitting a record $4,986 per capita, contrasting with stagnant levels of $163 in Sub-Saharan Africa.
        - Since 2015, global manufacturing employment has fluctuated. Starting at 14.3% in 2015, it dipped to 14.2% in 2020 but saw a marginal recovery in 2021. However, by 2022, it declined to 14.1%, with notable regional disparities. While Sub-Saharan Africa and Central and Southern Asia saw slight increases, other regions experienced declines, with Australia and New Zealand recording the highest fall of 0.5 percentage points from 2021 to 2022. These trends highlight the ongoing impact of crises on manufacturing employment growth.
    - **Target 9.3** Small enterprises face heightened vulnerability, particularly in low- income countries. According to survey data spanning from 2006 to 2023, only 16.9% of small-scale manufacturing industries in Sub-Saharan Africa had access to loans or lines of credit, compared to 45.4% in Latin America and the Caribbean. This underscores

how global uncertainty hampers investment flow and financial access for small businesses, hindering their resilience and adoption of new technologies.

- **Target 9.4**: Globally, CO_2 emissions per unit of GDP have steadily declined by 11.5% from 2015 to 2021, with a reduction of 16% observed in the manufacturing sector. Despite these positive trends, global CO_2 emissions from fuel combustion hit a record high of 33.6 gigatons in 2021, with manufacturing emissions also reaching their highest level since 2014 at 6.1 gigatons. These figures highlight the insufficient rate of reduction in CO_2 emissions intensity to achieve a significant overall decrease in worldwide CO_2 emissions.
- **Target 9.5**: After a slowdown in 2020, global research and development (R&D) expenditure appeared to return to pre-pandemic levels in 2021, climbing from 1.72% of GDP in 2015 to 1.93% in 2021. However, many developing economies have R&D expenditure relative to GDP below 1%.
- **Target 9.5**: The number of researchers per million inhabitants globally increased from 1143 in 2015 to 1352 in 2021, with Europe, Northern America, Australia, and New Zealand employing three times higher than the global level (i.e., 4050 and 4696 respectively in 2021). On the other hand, Sub-Saharan Africa has been substantially lower, standing at 96 researchers per million inhabitants. Additionally, women remain underrepresented, comprising only 31.5% of all researchers worldwide in 2021.
- **Target 9.b**: The gradual upward trend of medium and high-tech manufacturing value added in total value added, rising from 46.3% in 2015 to 46.9% in 2019, was briefly interrupted by the COVID-19 pandemic in 2020, dropping to 46.8%. Despite uncertainties and economic challenges, the sector displayed resilience with a slight decline of 0.67 percentage points in 2021. In Eastern and South- Eastern Asia, this sector accounted for approximately 50.6% of total manufacturing in 2021, whereas in Sub-Saharan Africa, it represented just 18.3%.
- **Target 9.c**: Mobile broadband (3G or higher) is accessible to 95% of the world's population, up from 78% in 2015. In most developing countries, this type of access is the main way—and often the only way—to connect to the internet. However, addressing the remaining 5% coverage gap poses challenges. Notably, in Oceania (excluding Australia and New Zealand), the gap remains significant at 31%. Although the gap in Sub-Saharan Africa is shrinking, it still stands relatively high at 17%, particularly impacting Central and Western Africa.

Goal 10. Reduce Inequality Within and Among Countries

27. Incomes of the poorest 40% of the population have generally grown faster than the national average in many countries and financial transfers during the pandemic boosted shared prosperity. However, over the past five years, the gap in per capita income growth between the poorest and richest countries

has widened. In addition, discrimination based on age, gender, religion, race, or belief affects one in six people globally. The year 2023 marked a record high of 35.8 million refugees, and over 8,000 migrant deaths were recorded globally. Addressing both within- and between-country inequality necessitates equitable resource distribution, investment in education and skills development, implementation of social protection measures, combating discrimination, supporting marginalized groups, and fostering international cooperation for fair trade and financial systems.

- **Target 10.1**: Among 124 countries with available data, more than half have achieved income growth for the bottom 40% of the population at a rate higher than the national average. However, there are notable regional disparities. In 78% of countries in Northern America and Europe, the bottom 40% experienced faster income growth than the national average, while only 30% of countries in Central Asia and Southern Asia demonstrated this trend. Limited data from the pandemic period suggests that in most regions, financial transfers boosted shared prosperity in many countries by supporting the income growth of disadvantaged populations.
- **Target 10.3**: The number of countries reporting on discrimination has increased by 37% since 2022. However, one person in six continues to encounter discrimination. Racial discrimination and discrimination based on age, gender, religion or belief remain pervasive. While 7% of the population surveyed report being discriminated on the ground of social origin or socioeconomic status, only less than a fifth of countries monitor this ground.
- **Target 10.4**: The share of economic output earned by workers decreased from 54.1% in 2004 to 52.7% in 2021, amounting to an average decline of $568 (PPP) per worker. The pandemic exacerbated this situation, with economic output and labor income in 2021 still below 2019 levels in many regions. As earnings from work are crucial for the less well-off and vulnerable, the long-term decline in labor income share represents an upward pressure for inequality.
- **Target 10.5**: In 2022, banks improved their overall performance compared to 2015 amid ongoing COVID-19 recovery. The proportion of countries reporting a return on assets (ROA) above one percent increased to 77.2% from 70% in 2021, with the median ROA rising from 1.34% to 1.56%. Asset quality also improved, with the median nonperforming loans to total loans ratio decreasing from 4.07% in 2021 to 3.52% in 2022. However, the capital buffer remained stable, with the median Tier 1 capital to risk-weighted assets at 16.8% in 2022, compared to 17% in 2021.
- **Target 10.7**: In 2023, there were 8177 migrant fatalities globally, marking the deadliest year on record, per data from IOM's Missing Migrants Project. This underscores the urgent need for safe migration pathways, as people continue to risk their lives on irregular routes due to limited alternatives.

- **Target 10.7**: At mid-2023, 35.8 million refugees under UNHCR's mandate, including other people in need of international protection, remained forcibly displaced due to war, conflict, persecution, human rights violations, and events seriously disturbing public order. This figure represents the highest total recorded to date and reflects increases due to new situations and no progress in resolving protracted refugee situations. In 2015, there were 213 refugees per 100,000 people worldwide, but by mid-2023, this figure doubled to 441 refugees per 100,000 people.
- **Target 10.c**: The global average remittance costs as a proportion of the amount remitted dropped from 7.42% in 2016 to 6.18% in 2023. Corridors offering costs below 5% increased from 23% in 2016 to 75% in 2023. While progress is evident, sustained efforts are needed to meet the target 10.c.

Goal 11. Make Cities and Human Settlements Inclusive, Safe, Resilient, and Sustainable

28. More than half the world's population currently reside in cities. However, cities are grappling with a multitude of complex issues, made more difficult by rising global urban poverty levels in the wake of COVID-19. From rising slum populations, insufficient public transport, city expansion outpacing population growth to threats to critical infrastructure, and disruption of basic services by disasters, it is essential that cities are equipped to adequately handle these challenges. As the world turns more urban, with nearly 70% of the global population projected to reside in cities by 2050, critical infrastructure, affordable housing, efficient transport, and essential social services are crucial for creating resilient, sustainable cities for all.
    - **Target 11.1**: In 2022, 24.8% of the urban population lived in slums or slum-like conditions, slightly lower than 25% in 2015, but higher than 24.2% in 2020. The total number of slum dwellers was 1.12 billion in 2022, 130 million more than in 2015. Over 85% of slum dwellers were concentrated in Central and Southern Asia (334 million), Eastern and South-Eastern Asia (362 million), and sub- Saharan Africa (265 million). Sub-Saharan Africa has the highest percentage of urban population living in slums, at 53.6%. Projections suggest that Sub-Saharan Africa will experience the greatest proportional increase in slum dwellers, with an additional 360 million expected by 2030. This emphasizes the urgent need for a comprehensive approach to address the urban housing crisis, including providing varied housing options and equitable access to basic services.
    - **Target 11.2**: Data collected in 2023 from 2039 cities across 188 countries shows that six out of ten urban residents globally have convenient public transport access, with notable gaps between developed and developing regions. In LDCs, only four out of ten people have access, compared to eight out of ten in more developed regions. Urgent investments are needed to expand access, especially in impoverished urban areas in developed countries.

- **Target 11.3**: Data from 1217 cities across 185 countries shows that cities are sprawling faster than they are densifying. Between 2000 and 2020, cities expanded up to 3.7 times faster than they densified. Globally, sprawl averaged 5.6% annually, while densification was only 1.5%. Despite a slight decline in sprawl rates from 2010 to 2020, they still exceeded densification rates, displacing ecologically valuable lands. Balancing people, prosperity, and the planet by 2030 requires concerted efforts to curb urban sprawl locally, sub-nationally, and nationally.
- **Target 11.5**: On average, 104,049 critical infrastructure units and facilities were destroyed or damaged by disasters annually from 2015 to 2022. Furthermore, disasters disrupted over 1.6 million basic services, including educational and health services, each year.
- **Target 11.6**: A comparison of air pollution five-year average before and after the development of the SDGs showed a significant decrease of 9% in fine particulate matter global levels and current alignment with the WHO Air Quality GuidelineInterim Target 1 value of 35 $\mu g/m^3$.
- **Target 11.7**: Data from 1365 cities across 187 countries reveals that access to open public spaces is notably deficient in LDCs, where fewer than three in ten people can conveniently reach such areas. Conversely, in high-performing regions like Australia and New Zealand, North America, and Europe, approximately six to seven out of ten urban residents enjoy convenient access to open public spaces, highlighting the prevalent global challenge.
- **Target 11.b**: In 2023, local-level risk governance has improved in recent years, with 106 countries cumulatively reporting having local disaster risk reduction strategies in place and in line with national strategies. On average, 72% of the local governments in reporting countries have specified having local disaster risk reduction strategies.

Goal 12. Ensure Sustainable Consumption and Production Patterns

29. The crisis of unsustainable consumption and production patterns worldwide is fueling the ongoing triple planetary crisis of climate change, nature loss, and pollution. Domestic material consumption and material footprint continue to rise, some one billion meals worth of edible food are wasted every day in homes around the world and stockpiles of e-waste steadily grow. While countries are fulfilling their environmental agreement obligations and embracing comprehensive approaches to address environmental degradation, public funding supporting the production and consumption of fossil fuels has more than tripled since 2015, impeding the transition to net-zero emissions. Each stage of production or manufacturing presents an opportunity to reduce resource and fossil fuel use, foster innovation, conserve energy, cut emissions, and advocate for a circular economy approach.
    - **Target 12.1**: From 2019 to 2023, one-third of member states (63 countries) have reported 516 policy instruments related to sustainable consumption and production.

- **Target 8.4/12.2**: From 2015 to 2022, Domestic Material Consumption (DMC) increased by 5.8%, and material footprint (MF) rose by 6.8%. Regional disparities between DMC and MF continue to grow, particularly between regions where MF is higher than DMC (Eastern and South-Eastern Asia, Europe and Northern America, Northern Africa, and Western Asia) and those where MF is lower than DMC (Central and Southern Asia, Latin America and the Caribbean, Sub-Saharan Africa, Oceania), showing different patterns of material consumption and their corresponding environmental impact.
- **Target 12.3**: In 2022, 19% of global food was wasted, totaling 1.05 billion tones, with household waste accounting for 60%. This waste generates significant greenhouse gas emissions, costing over $1 trillion annually, while 783 million people suffer from hunger. Addressing this issue is crucial for halving food waste by 2030, yet only nine out of 193 countries have included food waste in their Nationally Determined Contributions as of 2022. Meanwhile, the percentage of food lost globally after harvest on farm, transport, storage, wholesale, and processing levels is estimated at 13.2% in 2021.
- **Target 12.4**: Number of parties to international multilateral environmental agreements on hazardous waste, and other chemicals that meet their commitments and obligations in transmitting information as required by each relevant agreement:
  - Minamata Convention on Mercury: Most Parties have met their obligations, with 94% appointing National Focal Points and 95% submitting complete national reports in 2023.
  - Basel, Rotterdam, and Stockholm Conventions: Parties continue to make strides in meeting their obligations and in policy-making to keep pace with rapidly changing global circumstances, such as responding to the growing challenges created by contemporary waste streams such as plastic waste and e-waste, improving the procedure to control transboundary movements of wastes, and enlarging their scope with the listing of new chemical.
  - Montreal Protocol: Parties consistently fulfill reporting obligations on the production and use of ozone-depleting substances, with 156 out of 198 ratifying the Kigali Amendment by January 2024, demonstrating an increased commitment to mitigating climate change under the Protocol.
- **Targets 12.4 and 12.5**: In 2022, e-waste generation rose to 7.8 kg per capita from 6.2 kg per capita in 2015, but only 1.7 kg per capita was properly managed. Mismanaged e-waste leads to resource loss, increased use of virgin resources, and environmental hazards, underscoring the urgency for improved and environmentally sound management.
- **Target 12.6**: In 2021–2022, 73% of companies included in the sample published sustainability reports, with the number of companies tripling since 2016. This growth was observed in all regions in 2022.

- **Target 12.c**: Fossil fuel subsidies hit a record high of $1.53 trillion in 2022, reversing the declining trend observed from 2012 to 2020. The post-COVID energy price surge inflated these subsidies, prompting some governments to introduce new support measures. Consequently, public funding for oil, coal, and gas production and consumption more than doubled from 2021 to 2022 and tripled since 2015, impeding progress toward net-zero transition.

Goal 13. Take Urgent Action to Combat Climate Change and Its Impacts

30. Climate records were shattered in 2023, with the world watching the climate crisis unfold in real time. Communities around the world are suffering the effects of extreme weather, which is destroying lives and livelihoods on a daily basis. The roadmap to limit the rise in global temperature to 1.5°C and avoid the worst of climate chaos cannot afford any delays, indecision, or half measures by the global community. It demands immediate action for drastic reductions in global greenhouse gas emissions in this decade and the achievement of net zero by 2050.
    - **Target 13.1**:
        - The number of disaster-related deaths and missing persons per 100,000 population (excluding COVID-19 deaths) has nearly halved from 1.62 in the decade 2005–2014 to 0.82 in 2013–2022. However, the absolute number remains high. Between 2013 and 2022, disasters worldwide claimed 42,553 mortalities each year. Further, the number of persons affected by disasters per 100,000 population has increased by over two-third, from 1169 in 2005–2014 to 1980 in 2013–2022.
        - In 2023, 129 countries reported the adoption and implementation of national disaster risk reduction strategies, increasing from 55 countries in 2015. Among these countries, 122 countries have reported promoting policy coherence and compliance with the SDGs and the Paris Agreement as a key element in the strategy.
    - **Target 13.2**: The year 2023 broke every single climate indicator and was the warmest year on record according to the World Meteorological Organization. Global temperatures rose to 1.45°C, dangerously close for the first time to the

      1.5°C lower limit of the Paris Agreement on climate change. Despite some reduction in greenhouse gas emissions in developed countries, concentrations of greenhouse gases reached record high observed levels in 2022 and real-time data in 2023 show greenhouse gases continuing to increase. Carbon dioxide levels are 150% above pre-industrial levels.
    - **Target 13.3**: A study in 2023 of more than 530 grade nine science and social science subject curricula found that 69% contained no reference to climate change and 66% made no mention of sustainability. However, three-quarters of countries reported they have plans to revise their curricula in the next three years to focus more on climate change and sustainability.

- **Target 13.a**: Climate finance, reported by Annex I Parties as support provided to developing countries, has increased at a compound rate of 5% from 2015 to 2020, amounting to $41 billion. Although there are a range of estimates and a lack of an agreed accounting methodology on the $100 billion per year goal, the goal was not yet met as of 2021. However, recent progress made in the provision and mobilization of climate finance amounted to $89.6 billion in 2021.

Goal 14. Conserve and Sustainably Use the Oceans, Seas, and Marine Resources for Sustainable Development

31. Oceans cover over 70% of the Earth's surface and play a crucial role in providing food and livelihoods for more than 3 billion people as well as combating the effects of climate change. Yet, alarming trends from declining fish stocks, marine pollution, ocean acidification, and habitat destruction threaten marine ecosystems and the livelihoods of coastal communities worldwide. Urgent action is needed to address these challenges and ensure the long-term health and sustainability of the ocean through sustainable fishing practices, marine conservation efforts, pollution reduction, and global cooperation to safeguard marine life and ecosystems for future generations.
    - **Target 14.3**: Ocean acidification is increasing and will continue to do so if carbon dioxide emissions do not stop rising. An increasing number of countries and stations (from 178 stations in 2021 to 638 in 2024) highlights the growing capacity of countries to observe the continued decline of ocean pH in the global ocean as well as the strong regional differences in the pace of change.
    - **Target 14.6**: Illegal, unreported, and unregulated (IUU) fishing threatens the social, economic, and environmental sustainability of global fisheries, hindering countries' abilities to manage their fisheries effectively. The first binding international agreement to specifically target IUU fishing, the Agreement of Port State Measures, now has 102 states covered under the Agreement (from 25 in 2016), covering 63% of the world's coastal states. States have made good overall progress with close to 75% scoring highly in their degree of implementation of relevant international instruments in 2022 compared to 70% in 2018.

Goal 15. Protect, Restore, and Promote Sustainable Use of Terrestrial Ecosystems, Sustainably Manage Forests, Combat Desertification, Halt and Reverse Land Degradation, and Halt Biodiversity Loss

32. SDG 15 underscores the critical importance of biodiversity as humanity's life-support system. Yet, the relentless depletion of forests, coupled with an alarming rate of species extinction and stagnation in safeguarding key biodiversity areas, jeopardizes the delicate balance of our ecosystems. To address the pressing global environmental challenges and crises, including climate change, biodiversity loss, and pollution, as well as desertification, land and

soil degradation, drought and deforestation, it is imperative to intensify efforts in fulfilling our global environmental and biodiversity commitments.

- **Targets 15.1 and 15.2**: The world's forest area continues to decline, albeit at a slightly reduced pace compared to previous decades. The proportion of forest cover to total land area decreased from 31.9% in 2000 to 31.2% in 2020 with agricultural expansion accounting for nearly 90% of global deforestation. However, there has been notable progress toward sustainable forest management, marked by an increase in the proportion of forests under management plans and within protected areas. Moreover, certified forest area, which had shown steady long-term growth, has experienced a significant decline in the last two years, attributed to suspension of certificates due to the conflict in Europe.
- **Targets 14.5, 15.1 and 15.4**: Global protected and conserved area coverage of marine, terrestrial, freshwater, and mountain key biodiversity areas showed substantial improvements prior to 2000, this growth has stagnated over the last two decades. Particularly worrying is Central, Southern and Western Asia, Northern Africa, and Oceania, where average protected and conserved area coverage of key biodiversity areas is less than 30%. Progress has been more positive in Northern America and Europe, Sub-Saharan Africa, Latin America and the Caribbean, and Australia and New Zealand, where over 40% of each key biodiversity area is now covered on average.
- **Target 15.5**: Species extinction risk continues to worsen, as evidenced by a 12% deterioration in the aggregate Red List Index between 2024 and 1993 (and 4% since 2015). The extinction risk of the world's amphibian species was recently comprehensively re-assessed, revealing that for amphibians, climate change impacts, habitat conversion and alien invasive fungal disease are the most severe drivers of increasing extinction risk.
- **Target 15.6**: Countries continue to make progress in ratifying and implementing access and benefit-sharing instruments. By the end of 2023, 75 countries (up from six countries in 2016) and 93 countries (up from 12 countries in 2015) had reported on their legislative, administrative, or policy measures under the Nagoya Protocol and the International Treaty on Plant Genetic Resources for Food and Agriculture, respectively.
- **Target 15.7.1/15.c.1**: Estimates available for the first time show that, globally, intercepted illegal wildlife trade as a proportion of all wildlife trade (legal and illegal) increased from 2017 onwards, reaching its highest levels during the COVID-19 pandemic in 2020–2021. It is estimated that wildlife seizures made up around 1.4 to 1.9% of global wildlife trade in 2020–2021.
- **Target 15.9**: In 2023, 90 countries implemented the international statistical standard to measure the environment and ecosystems and their connection to the economy, an increase of 30% since 2017.

Goal 16. Promote Peaceful and Inclusive Societies for Sustainable Development, Provide Access to Justice for All, and Build Effective, Accountable, and Inclusive Institutions at All Levels

33. Around the world, we need peace in every sense. Yet the road to global peace and security has become even more complex. Global forced displacement is at an all-time high, exceeding 110 million people forced to flee from persecution, conflict, violence, and human rights violations as of May 2023. Between 2022 and 2023, civilian casualties experienced the highest spike since the adoption of the 2030 Agenda. Persistent threats to human security across the globe underscore the breakdown of peaceful and inclusive societies, crucial for sustainable development, and the need to restore trust and strengthen and renew global peace and security frameworks to heed the desperate pleas for peace from countless voices worldwide.
    - **Target 16.1**:
        - The global homicide rate gradually decreased, falling from 5.9 victims per 100,000 population in 2015 to 5.5 in 2020. However, this trend was disrupted in 2021, with a sharp rise to 5.8 victims per 100,000 population, only slightly decreasing to 5.6 in 2022. High levels of violence linked to organized crime and gang activities in Latin America and the Caribbean, along with Africa's heightened vulnerability to homicide, continue to contribute to these regions having the highest homicide rates and numbers globally.
        - The number of civilian deaths in armed conflict skyrocketed in 2023. Between 2022 and 2023, civilian casualties increased by 72%, the highest increase since 2015. In 2023, seven out of ten recorded deaths occurred in the Occupied Palestinian Territory and Israel. Moreover, for the second consecutive year, the number of civilians killed in conflict has risen dramatically; reversing the downward trend between 2016 and 2019. By 2023, the number of civilian casualties had risen to over 33,400, almost matching the peak of 2015. By 2023, four out of every ten civilians killed in conflicts were women, and three in ten civilians killed were children, doubling and tripling, respectively, the previous year's proportion.
        - Data on the experience of violence remain limited outside of Europe, Northern America, Latin America, and the Caribbean. Available data show women and men are not impacted by the same type of violence. The median prevalence of sexual violence in countries with data is 3.0% for women compared to just 0.8% for men. However, the median prevalence of physical violence is 3.0% for men compared to 2.1% for women.
    - **Target 16.2**: Violent discipline is the most common and widespread form of violence against children. In 82 (mostly low- and middle-income) countries with available data from 2015 to 2023, nearly eight in ten children from 1–14 years of age were subjected to some form of psychological aggression and/or physical punishment at home in the past month.

- **Target 16.3**:
    - Available data for 53 countries for the period 2010–2022 shows that the proportion of victims of physical assault and robbery in the previous 12 months, that reported their victimization to competent authorities, shows a median proportion in countries with data of 36% and 45%, respectively. This is notably higher than the proportion of victims of sexual assault that reported their victimization, for which the median in countries with available data was 17%.
    - Globally, the estimated number of persons in detention was 11.4 million in 2022, representing a prison population rate of 142 prisoners per 100,000 population. In 2022, nearly a third (3.5 million) of the global prison population was being held in pre-trial detention and their share remained stable between 2015 and 2022 (around 30%).
- **Target 16.5**: Based on data from 138 countries between 2015 and 2022, 19% of people who had contact with public officials in the last 12 months reported being asked to pay or paid a bribe to a public official. Regional differences range from an average of 32% in low-income countries to 9% in high-income countries.
- **Target 16.6**: Budget reliability improved in 2021 and 2022, reaching an average of 13.5% compared with 15.3% in 2020 but remained weaker than pre-pandemic levels with 10.6% on average. In part, this is due to the incidence of new international challenges including global political stresses, inflation, and resource price volatility.
- **Target 16.7**: The overall representation of people under the age of 45 or less in parliaments is increasing globally but with contrasting trends: a rise in developing countries and a decline in developed ones. Moreover, growth in women's share of parliamentary leadership posts continues to be slow-moving, albeit steady. As of January 1, 2024, women presided over 23.8% of parliaments as Speakers (an increase of 2.9 percentage points since 2021) and held 27.2% of committee chair posts.
- **Target 16.9**: While a few regions like Northern America, Europe, Australia, and New Zealand have achieved universal birth registration, only half of African children under five have had their births registered.
- **Target 16.10**:
    - The number of killings of human rights defenders, journalists, and trade unionists recorded by National Human Rights Institutions (NHRIs) and the United Nations decreased in 2023; to 320 cases in 40 countries, compared to 448 cases in 36 countries in 2022. In conflict zones, however, there was a sharp increase in journalists and media workers killed—40 lives lost—reversing the downward trend since 2017. Additionally, enforced disappearances have been nearly doubling for the second consecutive year, with at least 54 cases reported across 14 countries in 2023. Strong protection frameworks are therefore needed to stem this trend, particularly in countries where violent conflict or social unrest erupts or escalates.

    - In 2024, 140 countries had laws that specifically guaranteed the rights of citizens to access public information, up from 105 in 2015 and 14 in 1990.
- **Target 16.a**: Between 2015 and 2023, the number of countries with independent NHRIs meeting international standards increased by 23%. More than 40% of countries now have independent NHRIs.

Goal 17. Strengthen the Means of Implementation and Revitalize the Global Partnership for Sustainable Development

34. Global partnerships for sustainable development encompass key areas such as finance, technology, trade, and data. There are mixed trends in mobilizing financial resources for development, expanding internet connectivity, and strengthening statistical systems. However, a substantial $4 trillion annual investment gap for developing countries to achieve the SDGs, persistent and crippling issues such as unprecedentedly high external debt levels, and limited access to online connectivity in low-income countries underscore the need for sustained collaboration and enhanced cooperation and support in a landscape of worsening international cooperation and geopolitical tensions.

Finance

- **Target 17.1**: In 2022, the data from approximately 130 countries show that globally, government revenue accounts for approximately 33% of GDP. The average overall tax burden or revenue in the form of taxes is 26% of GDP among advanced economies, and 18% of GDP among emerging market and developing economies. In 2019, the overall average of proportion government expenditure funded by taxes was about 66% among advanced economies and 61% among emerging market and developing economies. The overall average sharply declined following the pandemic to about 52% in 2020 but rebounded in 2021 and 2022 for both groups of economies (to 62% for advanced economies and 59% for emerging and developing countries), however, still lower than the pre-pandemic level.
- **Target 17.2**: In 2023, ODA by member countries of the Development Assistance Committee (DAC) amounted to $223.7 billion, representing 0.37% of DAC members' combined GNI. Total ODA in 2023 rose by 1.8% in real terms compared to 2022 and by 47% compared to 2015. This was the fifth consecutive year ODA reached a new high. The increase was primarily due to aid for Ukraine, humanitarian aid, and contributions to international organizations.
- **Target 17.3**:
    - In 2022, financial resources for developing countries from multiple sources reported by 101 bilateral and multilateral providers amounted to $276.6 billion in official resources, $55.3 billion mobilized from private finance and $10.2 billion from private grants for development. Sustainable development grants (both official and private) decreased in 2022, compared to 2021. However, sustainable concessional development loans

increased by 6%, while non-concessional loans decreased and mobilized private finance increased by 21%, compensating the decrease of 2021.

- Global Foreign Direct Investment (FDI) flows in 2023 amounted to an estimated $1.37 trillion, a marginal increase over 2022. However, the increase was due largely to higher values in a small number of conduit economies; excluding these conduits, global FDI flows were 18% lower. The number of international investment projects announced in developing countries in sectors relevant to the SDGs—including infrastructure, renewables, water and sanitation, food security, health, and education—remained flat.
- The annual SDG investment gap in developing countries is now about $4 trillion. If the SDG investment needs to 2030 are to be met, some $30 trillion of additional investment must be found over the next eight years. More than half of the gap, or $2.2 trillion, relates to the energy transition alone.
- In the post-COVID period, remittances have proved to be resilient and become a premier source of external finance for developing countries. In 2022, remittance flows to low- and middle-income countries increased by 8%, to reach $647 billion. This increase is remarkable, given that it followed a 10.6% growth rate in 2021. The remittance growth rate is expected to moderate to about 4% in 2023.

- **Target 17.4**: The external debt stock level of low- and middle-income countries decreased in 2022 for the first time since 2015, to $9.0 trillion in 2022 from $9.3 trillion in 2021. Despite the slight decrease in 2022, external debt stock levels remained unprecedentedly high following more than a decade of rapid debt accumulation. Moreover, going forward, interest costs both in nominal terms and in relation to GNI and export revenue are expected to increase given the aggressive rise in global interest rates to tame inflation and could become increasingly burdensome by crowding out spending on other priorities for many low- and middle-income countries.
- **Target 17.5**: The number of countries that actively promote outward FDI to developing countries, including LDCs, remains limited. In 2023, at least 50 countries, including 19 emerging or developing economies, had at least one type of investment promotion mechanism for outward FDI in place. However, out of those, only 23 countries have adopted an outward FDI promotion scheme specifically targeting developing countries, including LDCs.

Information and Communications Technology

- **Target 17.6**: Fixed-broadband subscriptions continue to grow steadily, at an average annual growth rate of 6.4% between 2015 and 2023, reaching 19 subscriptions per 100 inhabitants in 2023 globally. Nevertheless, while fixed connections are common among households in upper-middle-income and high-income countries, they are nearly non-existent in low-income countries due to high prices and a lack of infrastructure.

- **Target 17.8**: Approximately 67% of the world's population, or 5.4 billion people were online in 2023. This represents a growth of 4.7% since 2022, a higher increase than that recorded from 2021 to 2022 at 3.5%. While there was an uptick in the increase in the number of internet users during the COVID-19 pandemic, in the last three years growth rates in the number of internet users were back to pre-pandemic levels.

Data, Monitoring and Accountability

- **Target 17.18**:
  - One of the far-reaching effects of the COVID-19 pandemic was the limited ability of national statistical offices to collect recent data for the Sustainable Development Goals. This was reflected in a drop in average data coverage scores in the Open Data Inventory (ODIN). Despite the recent decrease in data production capabilities, a comparison of ODIN coverage scores from 2017 to 2022 shows that the scores of low- and middle-income countries have increased at the same pace as high-income countries.
  - Globally, scores on the Data Sources Performance Index (Statistical Performance Indicators Pillar 4) and Data Infrastructure Performance Index (Statistical Performance Indicators Pillar 5) have been improving since 2016. Data sources improved by only three points, held back in part because of COVID disruptions, while data infrastructure—meaning both the hard and soft infrastructure needed to produce data are available—has increased by around 14 points.
  - In 2023, 159 countries and territories reported having national statistical legislation in compliance with the Fundamental Principles of Official Statistics, representing a significant increase from 132 in 2019 and marking the fastest annual growth of ten countries.
  - In 2023, a total of 163 countries and territories reported having implemented a national statistical plan, marking an increase from 143 in 2019 and 156 in 2022. Of these, 109 plans were fully funded, up from 91 in 2019 and 100 in 2022. These trends suggest a recovery from the long-term disruptions caused by the pandemic on the planning and execution of statistical activities.
- **Target 17.19**: There has been a resurgence in international support for the development of data and statistics, reaching $799 million in 2021, and representing a 14% increase from 2020 and a substantial 44% increase from 2015. Notably, 2021 marked the first time that multilateral aid providers emerged as the main source of funding.

Index

Pages in *italics* refer to figures.

I

J

K

L

M

N

O

P

Q

R

S